THE GEOLOGY OF BRITAIN

AN INTRODUCTION

PETER TOGHILL

Airlife

DEDICATION

Susan – for company in wild and often wet places, when
another photograph just needs to be taken.

Marianne and Elizabeth – for support and encouragement.

First published in 2000 by Swan Hill Press

This edition published by Airlife Publishing, an imprint of
The Crowood Press Ltd, Ramsbury, Marlborough, Wiltshire SN8 2HR

www.crowood.com

This impression 2019

British Library Cataloguing-in-Publication Data
A catalogue record for this book is available from the British Library.

ISBN 978 1 84037 404 9

The information in this book is true and complete to the best of our knowledge.
All recommendations are made without any guarantee on the part of the
Publisher, who also disclaims any liability incurred in connection with the use
of this data or specific details.

The front cover illustration is reproduced by permission of Ordnance Survey on
behalf of The Controller of Her Majesty's Stationery Office, © Crown
Copyright MC 100038003.

Typeset by Rowland Phototypesetting Ltd, Bury St Edmunds

Printed and bound in India by Replika Press Pvt. Ltd

THE
GEOLOGY OF BRITAIN

AN INTRODUCTION

ACKNOWLEDGEMENTS

I must sincerely thank Gill and Andrew Jenkinson of Scenesetters, Bucknell, Shropshire, who have produced all the computer generated colour diagrams. Gill Jenkinson in particular has worked with skill and patience when deciphering my original drawings, and has put up with many changes of opinion. The original manuscript was all hand-written, and I owe heartfelt thanks to my secretary Brenda Campbell who has word-processed the manuscript and is one of the few people who can read my handwriting. She has also shown great patience and coped with numerous versions of various sections.

Thirty (nearly half) of the photographs were generously made available, as indicated, by Ken Gardner of Landform Slides, Lowestoft. Without his support the book would not have been so attractive.

Dr Andy Chambers, School of Earth Sciences, The University of Birmingham, kindly provided the photograph of layered gabbro on Skye (fig. 149). A number of publishers have allowed me to re-draw diagrams from their publications as indicated in the text, and I am most grateful for this. Duplicate sets of photographs made available by Landform Slides can be purchased from Mr K. Gardner, Landform Slides, 38 Borrow Road, Lowestoft, Suffolk N32 3PN.

FOREWORD

The geology of Britain is immensely varied, with rocks and structures representing over 2000 million years of earth history. Geologists working in Britain were amongst the first to recognise the importance of fossils in working out the relative ages of rocks. The fossils discovered in Britain during the nineteenth century paved the way for the setting up of a geological time scale now recognised worldwide. The varied scenery of Britain is wholly dependent on the underlying geology and geological structure.

This book sets out to describe the geological history of Britain through the various recognised periods of geological time, starting with the oldest rocks in the north-west of Scotland and finishing with the young, unconsolidated sediments of south-east England, and those left behind by the recent Ice Age.

There is a public interest nowadays in geology, particularly earthquakes and volcanoes, as explained by the modern concept of plate tectonics. This book uses plate tectonics to explain the geological development of Britain and its remarkable movement across the Earth's surface in the last 500 million years. It is written for the lay person and includes an introduction to general principles of geology, but will also appeal to students at all levels.

The reader will find a geological map of Britain a very useful complement to the book. The most suitable is the British Geological Survey's Ten Mile Map, Scale 1: 625 000, Solid Edition. This comes in two halves (north and south), and each can be purchased through stockists of Ordnance Survey maps, or direct from the British Geological Survey (see References for the address).

CONTENTS

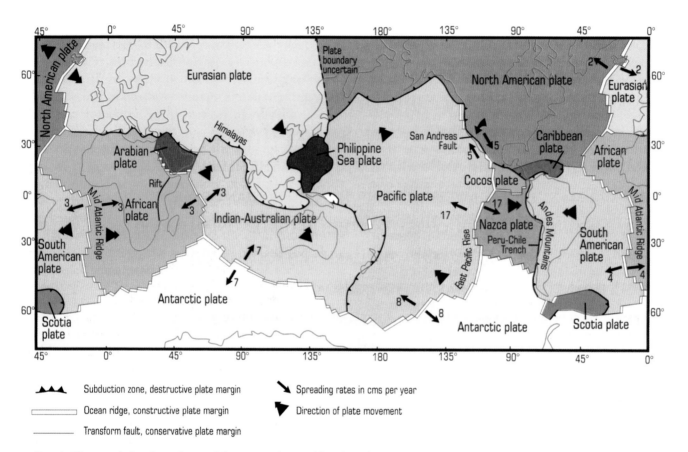

FIG. 1. *Plates and plate boundaries of the present day world. After Plummer and McGeary,* Physical Geology, Brown, W. C., *Times Mirror Education Group 1996, 7th edition, p. 418, fig. 19.1.* (Reproduced with permission of The McGraw-Hill Companies.)

INTRODUCTION

Visitors to Britain are always struck by the great variety of scenery in what is a relatively small geographical area. This ever-changing scenery and landscape pattern is a reflection of an underlying complex geological sequence and structure developed through hundreds of millions of years of Earth history.

One of the reasons why the geological sequence is so varied is that the British area has often in the past (though happily not today!) been close to plate boundaries. These boundaries are the active geological areas of the world where constant earthquake and volcanic activity occurs, and where compression of the Earth's crust can build up large mountain chains. The presence in Britain of ancient fault lines such as the Great Glen Fault and old volcanic areas like Snowdonia, together with the evidence of deeply eroded fold mountains in Scotland dating back over 400 million years, all point to our having been at plate boundaries in the past.

The various rock types of Britain have been formed in a number of ways. During the hundreds of millions of years which have shaped the landscape, constant erosion of land masses produced sediments which were laid down to form new sedimentary rocks, such as sandstone and shale, of which many parts of Britain are made. Episodes of igneous activity have produced the volcanic lavas and ashes, and granites, which we find in many of Britain's highland areas, and these episodes have occurred at various times in the geological past. Many of the older rocks of Britain have been altered by the effects of heat and pressure caused by deep burial in the Earth's crust, and movements at plate boundaries. These altered rocks, which geologists call metamorphic, include slates and schists. Other rock types such as reef limestones, desert sandstone and coal-bearing sediments show evidence of the British area having been in very different latitudes from those of today. This suggests large movements of continental areas over the surface of the globe, which are explained by the modern concept of plate tectonics.

Plate tectonics

This concept explains how continents have moved over the Earth's surface throughout geological time, and are still moving today. The early ideas of continental drift, as put forward by Alfred Wegener, although basically correct, could not produce a sensible explanation or mechanism for continental movements. It was not until the discovery in the 1960s of ocean floor growth caused by volcanic activity, now called sea-floor spreading, that a mechanism for the movement of continents, as well as for the birth and destruction of oceans, was discovered.

Geologists now accept that the Earth's outer shell, down to around 100 km, is made up of a number of distinct areas called plates (fig. 1). These plates move independently of each other, driven by the earth's internal heat in the form of giant convection cells. The boundaries of the plates, where they are in contact with each other, are lines of intense movement (tectonic activity), where one plate is over-riding or sliding past another, or where new molten material, in the form of basalt lavas, is coming up to the Earth's surface (fig. 2). These latter boundaries, called constructive plate boundaries, occur at the now-famous mid-ocean ridges, where volcanic activity produces a constant conveyor belt of new oceanic crust on either side of the ridge to produce a new ocean floor. These new ocean floor 'factories' move continents apart to a certain extent on either side of the ocean, as is happening around the Atlantic today.

The constant supply of new ocean-floor material would make the Earth's surface area grow larger and larger if it were not for the fact that elsewhere ocean-crust material is dragged back into the Earth's mantle at destructive plate boundaries. These destructive boundaries take the form of deep ocean trenches which mark the movement of material back down into the mantle. This process is called subduction and a subduction zone inclines downwards towards the mantle, showing where one plate is descending below another. The area around a subduction zone is a site of intense earthquake and tectonic activity where rocks are folded, faulted, altered (metamorphosed) and melted to form new igneous rocks. The majority of the world's major earthquakes and volcanoes are associated

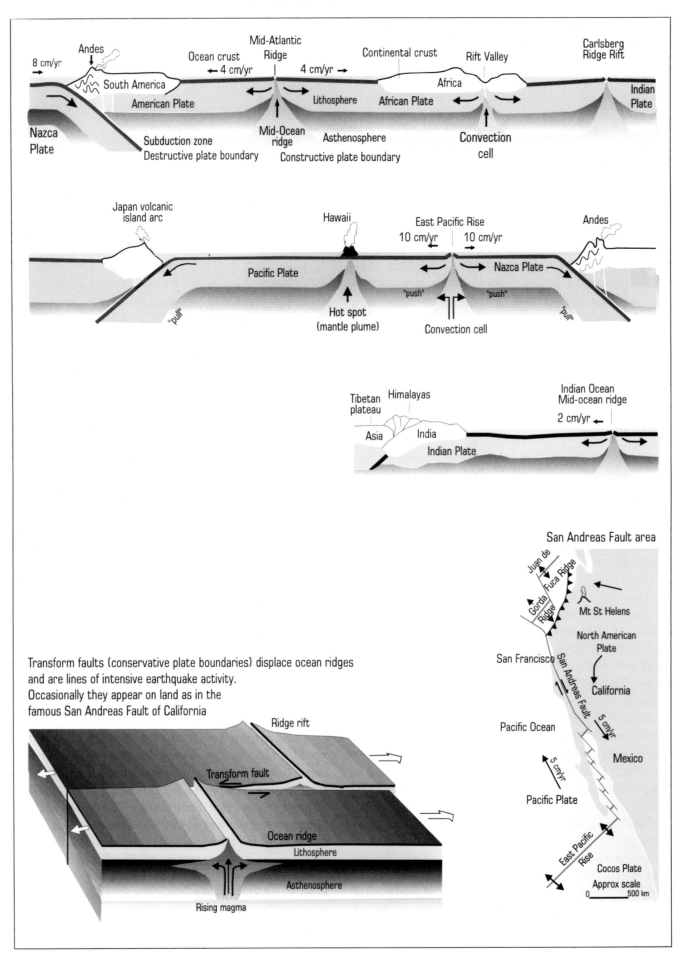

FIG. 2. *Plate tectonics explained using the present arrangement of continents and oceans. After Press and Siever,* Understanding Earth, *Freeman, 1998, 2nd edition p. 513, figs. 20.7 and 20.8.* (Reproduced with permission of the McGraw-Hill Companies.)

with subduction zones. However some of the most intense earthquake activity occurs where one plate is sliding past another along major breaks in the crust called transform faults (fig. 2). These plate boundaries are called conservative plate boundaries and the best known is the San Andreas Fault in California (fig. 2). Fold mountains and volcanoes are also formed on the leading edge of the plate, which is over-riding the other if it is of continental material. This is happening in the Andes of South America. Volcanic island arcs occur when the two colliding plates are both oceanic areas, as in the East Indies today.

Sea-floor spreading is certainly driven by convection cells causing a 'push' effect (fig. 2) at ocean ridges. But widening is also caused by a 'pull' effect as cold dense plates slide down subduction zones due to gravity. Finally the process of sea-floor spreading which causes new oceans to form and grow will also lead to the destruction of older oceans, and the end product of this is two continents on either side of an old ocean colliding to form a huge fold mountain chain as in the modern Himalayas (fig. 2).

Magnetic anomalies and sea-floor spreading

The great breakthrough in plate tectonics in the 1960s came about because of studies of the magnetic field of the ocean floor, and this led to the discovery of sea-floor spreading and a mechanism for continental drift. We now know that the Earth's magnetic field repeatedly reverses every few hundred thousand years or so, and there are thus large intervals of time (fig. 3) when the field was opposite to that of today – when a compass needle would have pointed to the south magnetic pole. We refer to the magnetic field as showing reversed or normal polarity, and we can trace these cycles of changes back for over 400 million years, although the last 250 million years are best documented.

When lavas are erupted from volcanoes the iron minerals, particularly in basalts, take up the magnetic polarity of the time (reversed or normal) and this is then frozen in them and not changed by later magnetic field reversals. A study of basalt lavas dated by radioactive methods on land going back over four million years has produced the magnetic reversal time scale shown in fig. 3. Studies of the ocean floor around mid-ocean ridges have shown them to be made entirely of basalt, and to have a striped pattern of magnetic anomalies symmetrical about the ridge axis (fig. 3). The magnetic anomalies are strengths of the field greater and less than average, and it is was shown by Vine and Matthews in 1963, following work by Hess in 1960, that these stripes corresponded to normal and reversed polarity in the basalts of the ocean floor. Hess had suggested that basaltic ocean crust was newly created at ocean ridges, and a comparison of the striped pattern on either side of the ridge axis on the ocean floor showed it to be identical to the reversed-normal pattern found in basalt lavas on land (fig. 3). Thus by calculating how far a particular stripe was from the ocean ridge, and knowing the age of this stripe from continental lavas, it was shown how far the stripe had moved in so many million years from the ridge it originated from (fig. 3). Thus the concept of sea-floor spreading was conceived and we can now calculate spreading rates all over the world by comparing the magnetic anomaly pattern on the sea floor, which represent normal-reversed polarity in the basalt crust, with the known age of the anomalies as calculated from continental basalts. Sea-floor spreading rates vary from 2 cm per year in the North Atlantic to 20 cm per year in the North Pacific. These are the total amounts produced by the two limbs on either side of the ocean ridge.

Sea-floor spreading is the mechanism by which oceans grow and widen, and push continents apart. However all oceanic crust is eventually destroyed at subduction zones, where it dives back down to the mantle. Hess suggested that mid-ocean ridges were ephemeral features having a life of 200–300 million years; and in fact nowhere on the Earth today is there any ocean crust older than 200 million years – it has all been destroyed. However, as we shall see later, evidence for old oceans up to 500 million years old comes from the marine oceanic sediments of older periods, which are not subducted, old volcanic island arcs and rare pieces of old oceanic crust, called ophiolites, which are preserved within continents. We can also show how continents have moved through various latitudes in the past from the ancient climates shown in the rock sequences, and these continents must have been pushed around by sea-floor spreading within oceans long since destroyed.

All the effects of plate tectonics can be seen happening somewhere on the Earth today, and where we see these effects in older rocks we can conclude that these areas were at plate boundaries in the past.

Thus, in examining the geology of Britain, we can see for instance the great thicknesses of volcanic lavas and ashes of Snowdonia which were formed between 500 and 450 million years ago. These are the product of a destructive plate margin where an oceanic plate was subducted under another to form a series of volcanic islands as material melted at depth. The volcanic ashes often fell into the sea and entombed animals living on the sea bed. These are now found as fossils within beds of ash and can be found on the very summit of Snowdon. A modern analogy is the volcanic island arc of the East Indies with the surrounding oceanic area teeming with life.

The modern science of geochemistry, which

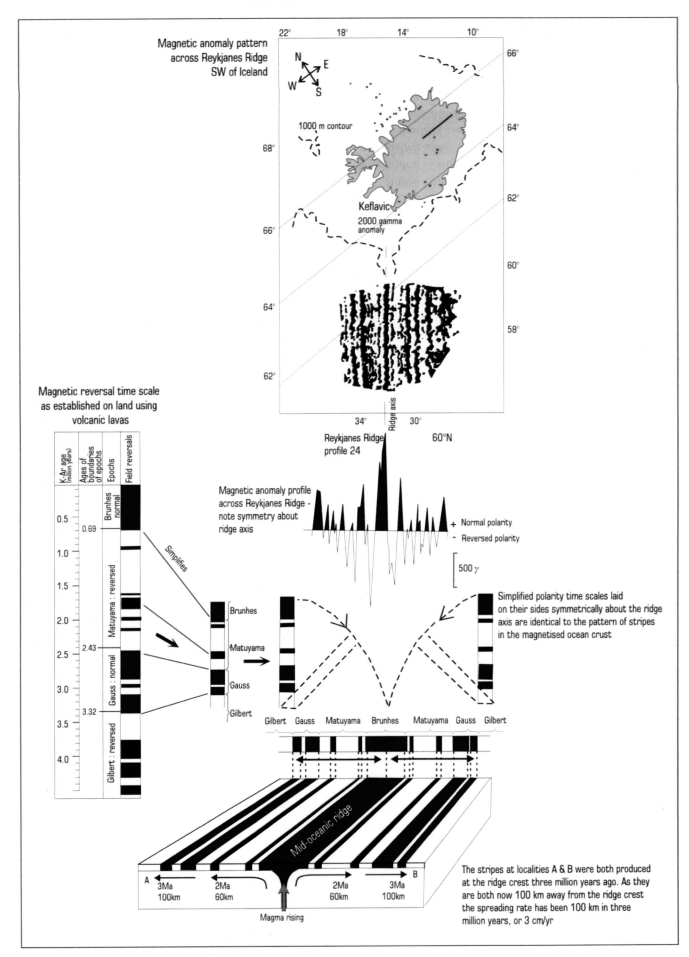

FIG. 3. *Magnetic anomalies and sea-floor spreading. After Open University Course 102, Blocks 7 & 8,* Plate Tectonics, *pp. 43, 45.* (Reproduced with permission of the McGraw-Hill Companies.)

analyses the chemistry of rocks, can identify trace element differences between volcanic rocks from continental areas such as the Andes and those from oceanic areas such as the East Indies. By applying this process to ancient volcanic rocks we can show that those from Snowdonia and the older lavas of Shropshire, the Wrekin, etc., are from island arcs whereas those from the Midland Valley of Scotland are of continental origin.

The Great Glen Fault in Scotland may mark the site of an old conservative plate boundary where one plate slid past another around 400–350 million years ago and caused enormous earthquakes. A modern analogy is the famous San Andreas Fault in California, where the Pacific plate is moving northwards, sliding past the North American plate and causing numerous severe earthquakes.

The old fold mountain chain which covers the Scottish Highlands area is an example of an ancient continental collision around 400 million years ago which destroyed an ancient ocean. The resulting collision folded up mountains up to 8000 m high and the process compares exactly with that which caused the Himalayas to form when India collided with Asia about 20 million years ago. Thus plate tectonics can now be applied to the study of older rocks as well as the Earth's present-day features. Geologists now accept that processes happening on and within the Earth today, have been happening in the same way for over 3000 million years.

The geological sequence in Britain shows evidence of rocks produced in a variety of climates from cold temperate conditions to sub-tropical reef environments, Sahara-like deserts and tropical rain forests. Instead of proposing widespread global climatic changes in the past we can now simply say that Britain has drifted northwards during the last 500 million years from a position in the southern hemisphere, with southern Britain near the Antarctic Circle, across the equator and into the northern hemisphere reaching its present latitude around 50 million years ago. Since then movement following the birth of the North Atlantic has been mainly eastwards. We probably spent 300 million of those 500 million years in tropical latitudes, which explains the coral reefs, coal forests and Sahara-type deserts.

This assumes that the Earth has usually had the same climatic conditions and temperature differences between the poles and the equator. However we now believe that during most of the last 1000 million years the Earth has been warmer than today, but that during four particular periods there has been major cooling, which produced Ice Ages. We are living in one of these cooler periods today, and it has affected Britain for about two million years. So today the Earth is cooler than usual; the normal situation in the past has been such that there has been no ice at the poles, leading to higher sea levels. During cold periods oceanic water is trapped in continental polar ice caps and this lowers sea levels.

Because of our position near to plate boundaries in the past, and often at different latitudes, the great variety of rock types and structures which has resulted allows us to use Britain to understand many fundamental principles of geology. We can also use plate tectonics to explain many past events.

Geological time

In the nineteenth century the biblical belief in a young, 4000-year-old, Earth was gradually superseded by the view, based on scientific discoveries, that the Earth was very old, in fact many millions of years old. The early concept of catastrophism, which assumed that every feature on the Earth's surface was formed by catastrophic action or biblical flood, was gradually replaced by the ideas of uniformitarianism, which suggested that the Earth's surface has been changed immeasurably slowly, and the processes which formed rocks and geological features in the past are still happening today at the same rate. In other words, 'the present is the key past' – we can explain all ancient features in terms of modern processes. In fact plate tectonics is just the latest in a sequence of uniformitarian theories.

These theories were first sensibly put forward by James Hutton in 1795 in his book *Theory of the Earth*. Hutton was a well-known Scottish philosopher, scientist and gentleman farmer, and a leading member of the Scottish Enlightenment. He studied many geological features around Edinburgh and southern Scotland and explained them in terms of igneous and sedimentary processes. His most famous locality is at Siccar Point on the Berwickshire coast, where he described what we now call an unconformity, a structural difference between two sedimentary rock sequences, usually a marked difference in dip (fig. 4 and 5), caused by a period of uplift and erosion between the formation of the two sequences. An unconformity usually represents a long interval of time, about 50 million years in the case of Siccar Point (fig. 4).

Here Hutton saw coarse sandstones of the Old Red Sandstone in near-horizontal layers overlying a deeply eroded sequence of hard grey Silurian sandstones with vertical layers or bedding (fig. 4). He clearly realised that one sequence had been laid down, hardened, folded up and eroded before the next was laid down on top. In looking at these exposures, he said in 1788 that he could see, in terms of Earth history, 'no vestige of a beginning and no prospect of an end'. This famous

FIG. 4. *Unconformity at Siccar Point, Berwickshire, first described by Hutton in 1795.* TOP PICTURE: *The classic view with gently dipping Upper Old Red Sandstone rocks resting on steeply dipping Silurian greywackes.* (Copyright, Landform Slides.) LEFT: *View from the south showing a thicker cover of Old Red Sandstone.*

sentence indicates Hutton's belief in an Earth of immeasurable age – an opinion which was supported in the nineteenth century by many geologists, as well as Charles Darwin.

Another famous example of an unconformity occurs in Shropshire. During the Ordovician period, 510–450 million years ago, a sea covered most of the county and laid down a sequence of marine sediments. During the late Ordovician the Shropshire area suffered compression and uplift due to plate tectonic processes, and these gave rise to a number of folds and faults. The folded rocks suffered erosion for maybe ten million years, which produced an uneven landscape. During the succeeding Silurian period around 440 million years ago, the sea spread again over the eroded area to lay down a series of horizontal sediments on top of the folded and eroded Ordovician landscape (fig. 5). If this boundary is exposed, one can actually see the later Silurian sediments resting on Ordovician rocks which are clearly folded and have a different inclination or

dip. This unconformity, although of a different age, is exactly the same as that described by Hutton at Siccar Point. In both cases we can see an angular discordance, or difference in dip, between the rock sequences of the two different periods. Another definition of an unconformity states that 'time is not represented by rocks'.

Unconformities occur throughout geological time, and throughout the world, and are indications of the constant interchange between land and sea, erosion and deposition. Another famous one occurs in the Pennines at Horton-in-Ribblesdale (chapter 7, fig. 92).

Fossils
Fossils have been known for hundreds of years, but their significance as indications of ancient life forms was not recognised until the nineteenth century. During this time fossils were systematically collected from many different areas of Britain and it became clear to Victorian geologists, and to Charles Darwin, that they represented an evolving sequence of life forms.

It was William Smith (often called the 'father of stratigraphy') who when working on canal construction in southern England in the early 1800s showed that each layer of rock contained a distinct set of fossils, and that strata could thus be recognised by

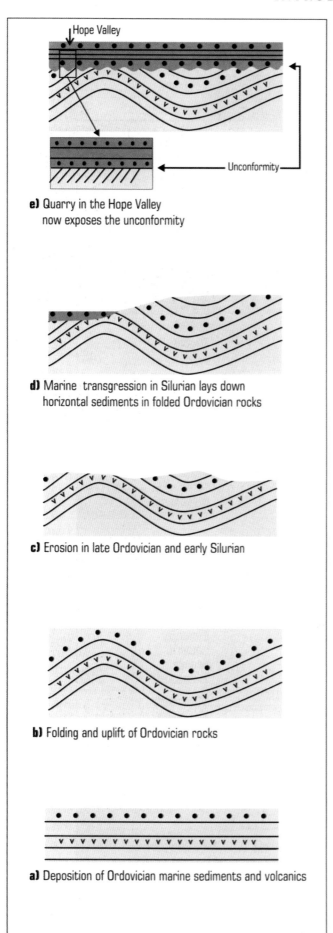

e) Quarry in the Hope Valley
now exposes the unconformity

d) Marine transgression in Silurian lays down
horizontal sediments in folded Ordovician rocks

c) Erosion in late Ordovician and early Silurian

b) Folding and uplift of Ordovician rocks

a) Deposition of Ordovician marine sediments and volcanics

FIG. 5. *Unconformity in Shropshire between the Ordovician and Silurian Systems.*

the fossils they contained. In fact it became possible to recognise different levels in the enormous rock sequence covering Britain by the fossils discovered in the 1800s. Stratigraphy, the study of the layered sequence of sedimentary rocks (strata) laid down throughout geological time, was pioneered in Britain by people like William Smith, who by 1815 had produced a geological map of England and Wales. Investigations over the whole of Britain led to a sequence of rocks being recognised and divided into major units called systems, each often thousands of metres thick, but each with a distinct set of fossils. A relative geological time scale was thus erected (fig. 6) that had the oldest rocks at the bottom of the pile of sediments and the youngest at the top. It must be remembered that this classification is based entirely on fossil content, with one group of fossils following another. The boundaries between systems are defined by major changes in fossil content, which often show up as extinction events nowadays believed by many people to have been due to asteroid or comet impacts with the Earth. The terms used for the three eras of geological time – Palaeozoic, Mesozoic, and Cainozoic – literally mean 'ancient', 'middle' and 'new life'.

When studying a great thickness of layered sedimentary rocks (fig. 7) it is assumed that those at the bottom of the sequence were deposited before those at the top. This is called the Law of Superposition. It sounds obvious but of course when rocks become folded as in fig. 7, it can be difficult to decide which are at the bottom of the sequence. In this situation, a knowledge of the fossils can provide the answer, as can sedimentary structures such as channel bedding.

In North Wales the Ordovician System of rocks (fig. 8) comprises up to 8000 m of sandstones, mudstones and other sedimentary rocks as well as lavas and ashes. These rocks contain a distinct set of fossils. (The term Ordovician comes from the name of the ancient British tribe who inhabited North Wales, where rocks of this age were first studied in the mid-1800s.) By studying the inclination of the rock layers it can be seen that these Ordovician rocks rest on top of another sequence or system, comprising thousands of metres of sediments, again with their own distinct set of fossils. This is called the Cambrian System, which was again first studied in Wales and named accordingly. If we go back to the top of the Ordovician, particularly in south-east Wales and the Welsh borders, we can see that another sequence, the Silurian System, named after the Silures tribe of south-east Wales and the borders, followed on top of the Ordovician, again with its distinctive fossils. We refer to the time in which a system was formed as a geological period, and thus the Silurian System was formed during the Silurian Period. We can continue

Eons	Eras		Periods / Systems		Epochs / Series	
Phanerozoic	Cainozoic		Quaternary		Holocene	10000
					Pleistocene	2
			Neogene	Tertiary	Pliocene Miocene	24
			Palaeogene		Oligocene Eocene Palaeocene	65
	Mesozoic		Cretaceous		Upper Lower	146
			Jurassic		Upper Middle Lower	208
			Triassic		Upper Middle Lower	245
	Palaeozoic	Upper	Permian		Upper Lower	290
			Carboniferous		Stephanian Westphalian Namurian Dinantian	363
			Devonian		Upper Middle Lower	409
		Lower	Silurian		Pridoli Ludlow Wenlock Llandovery	439
			Ordovician		Ashgill Caradoc Llandeilo Llanvirn Arenig Tremadoc	510
			Cambrian		Merioneth St David's Caerfai	544
Precambrian (Cryptozoic)	Proterozoic	Neo	1000	Precambrian	Dates in millions of years ago, Ma.	
		Meso	1600			
		Palaeo	2500			
	Archaean	Late	3000			
		Middle	3400			
		Early	3800			
	Pre-Archaean					
			4600 Ma	Formation of Earth		

FIG. 6. *Geological time scale.*

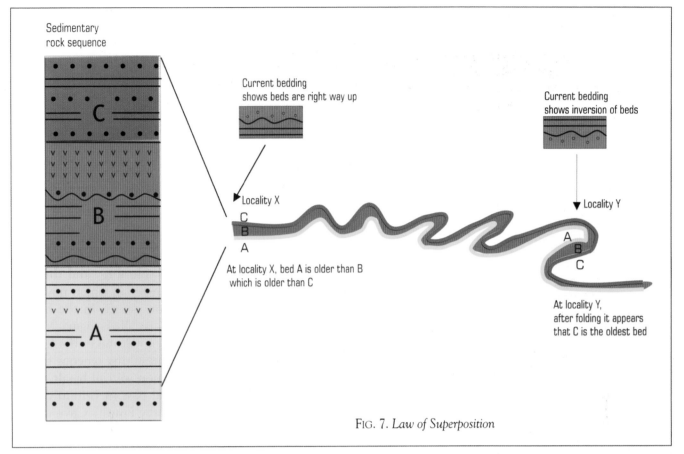

Sedimentary rock sequence

Current bedding shows beds are right way up

Current bedding shows inversion of beds

Locality X

C
B
A

At locality X, bed A is older than B which is older than C

Locality Y

A
B
C

At locality Y, after folding it appears that C is the oldest bed

FIG. 7. *Law of Superposition*

going up the stratigraphical column in England and Wales through the various geological systems until we arrive at the present day. We are living in the Quaternary Period, and rocks of the System are still being formed as it only started two million years ago. We thus have thirteen geological systems (fig. 6), although the first of these, the Precambrian, covers a vast interval of time, and also contains very few fossils. Precambrian rocks have to be subdivided in different ways, according to episodes of metamorphism, and radiometric dating (see below), for instance.

The stratigraphic sequence is thus a *relative* time scale, with each system following on from another. We can tell where we are in the sequence because of the fossils we find, but we cannot at this stage know how old the rocks are in millions of years.

We can move on to see how we can calculate *absolute* ages, and establish how old rocks actually are in terms of their formation millions of years ago. The science of geochronology, the absolute dating of rocks, did not appear until the early twentieth century when the slow radioactive decay of certain elements allowed the dating of mineral grains in rocks. In very simple terms, certain grains in igneous rocks contain, for example, a radioactive isotope of uranium which decays to give an isotope of lead. The rate of decay of this isotope is known – it is a constant which can be calculated in the laboratory. The decay is very slow; the half-life, as it is called, of uranium is over 4000 million years. If the amount of decay products in a rock can be measured,

and compared with the amount of original isotope remaining, then the period of time during which the mineral has been decaying, and hence the age of the rock can be calculated. This is called a radiometric date, and assumes that all the decay products have come from the parent isotope as soon as the mineral crystallised. The date obtained, say 500 million years, is given the abbreviation 500 Ma (millions of years ago), and from now on this is the term I shall use.

Radiometric dating can use many different elements as well as uranium/lead, and is nowadays a very sophisticated method of dating igneous rocks. It is also easy to apply this to sedimentary rocks and stratigraphy, since if a lava flow occurs at the end of the Ordovician Period, and just before the start of the Silurian, then the date of the lava flow gives us the date of the Ordovician–Silurian boundary. Over the last fifty years more and more rocks have been dated and the detailed time scale set out in fig. 6 has been worked out. This allows us to have two sets of nomenclature – lithostratigraphic, which relates to the rock stratigraphic sequence, and chronostratigraphic for the actual ages in millions of year ago. Thus the Cambrian System is a sequence of rocks laid down during the Cambrian Period between 544 Ma and 510 Ma. Smaller subdivisions refer to series of rocks; for example, the Wenlock Series was formed during the Wenlock Epoch, part of the Silurian Period. And within the Wenlock Series are individual rock units like the famous Much Wenlock Limestone.

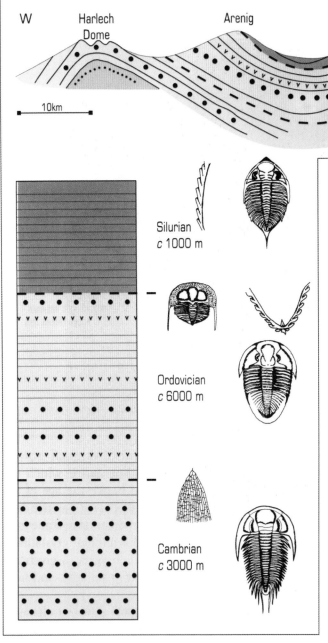

FIG. 8. *Cross-section through part of North Wales demonstrating the law of superposition and relative ages of rocks. Section is along C–D on fig. 29.*

The time scale set out in fig. 6 is that accepted by most geologists in the 1990s. As dating techniques improve it is bound to change. The latest change is the base of the Cambrian from 570 Ma to 544 Ma. However, major changes are unlikely. It must be emphasised that the decision as to whether a rock is, say, of Devonian age or Carboniferous, when it is close to the boundary, is made entirely on the basis of fossils, not absolute dating. The fixing of these system boundaries using fossils is a world-wide debate which has been going on for a hundred years.

Rock types

Before we can understand the geological history of

Britain we need to know the different types of rocks as defined by geologists.

Igneous

The very first rocks on the Earth's surface were formed from the cooling of molten solutions, and they are called igneous. The molten material, which nowadays comes from relatively shallow depths in the Earth's mantle – up to 60 km down – is called magma. Magma can come to the surface from a volcano in the form of lava, and when it cools the resulting igneous rock is often of a chemical composition we call basalt. Good examples from the geological past in Britain can be seen on Mull and in the nearby Fingal's Cave on the Isle of Staffa. Other types of lava include andesite (common in the Andes), and obsidian. Sometimes lava has already hardened in the volcanic vent and is then thrown out as volcanic ash during violent explosions. This ash, called tuff if it is from an ancient eruption, can entomb whole areas, as it did at Pompeii. It can also be erupted as a dense, incandescent, gas cloud which flows down the mountain at great speed and causes great damage. Such an eruption in Martinique in 1903 killed over 30 000 people, and the early stages of the eruption of Vesuvius in AD 79 were probably of this type.

Large areas of the Earth's crust can become molten near to subduction zones and the resulting magma rises up and cools in the crust to form a large mass of granite. This type of coarse grained igneous rock is called intrusive as it pushes its way through the crust, but never actually reaches the surface. Granites of the geological past are common in Britain, as on Dartmoor and in the Cairngorms. They formed at depth in the crust, and have been exposed today after millions of years of erosion. Smaller intrusions often reach higher in the crust and are called dykes and sills. The famous Whin Sill in the northern Pennines is a thick layer of basaltic rock forming a conspicuous north-facing cliff, which is followed for miles by Hadrian's Wall.

When describing igneous rocks geologists refer to them chemically as being either acid, intermediate or basic. This refers to the amount of silicon dioxide (SiO_2 or silica) in the rock, and our use of the terms acid and basic is in no way comparable to the terms acid and alkaline used in chemistry and soils, etc.,

FIG. 9. *Horizontal strata forming a 3500 m high peak, Mt Temple, in the Canadian Rockies near Lake Louise.*

which are based on hydrogen ion content and referred to as pH value.

Igneous rocks vary in silica content from acid – around 75 per cent, as in granites – through intermediate – 55 per cent, as in andesites – to basic – around 45 per cent, as in basalts. The change from acid to basic includes an increase in calcium silicates but even basalts would have an acid pH value.

We do not use the terms acid and basic for metamorphic or sedimentary rocks, as these can contain calcium carbonate – as in marbles and limestones – and have alkaline pH values.

Sedimentary

When areas of the Earth's crust exposed as land are eroded by the sea, or by rivers, wind, and ice, the material formed as small particles is called sediment. This can be transported away from the source and laid down in layers on the sea bed, in lakes or river estuaries, or on land, to give great thicknesses of sediment. Over long periods of time these are hardened to form sedimentary rocks, which usually have an obvious layered structure. Owing to plate tectonic processes these sedimentary rocks can be folded up and raised above sea level to form new continental areas. Some of the world's highest mountains are formed of sedimentary rocks, and in areas such as the Canadian Rockies the layered structure can be clearly seen (fig. 9).

Some of the most best-known sedimentary rocks are sandstones, which are made of relatively coarse particles; and shales, and mudstones, which are made of finer clay particles. Limestones are calcareous sedimentary rocks made up of fragments of the shells of dead marine animals and plants, and also a proportion of pure calcium carbonate precipitate. They sometimes include whole fossilised reefs – evidence of the area having been in subtropical latitudes.

Metamorphic

Heat and pressure close to plate boundaries can cause rocks to be altered both physically and chemically so that the new types appear very different from their parent rocks. When shales and mudstones are metamorphosed they turn initially to slates and then to coarse-grained foliated rocks rich in shiny mica, called schists, and finally to coarse-banded rocks called gneisses. Some of the world's oldest rocks, dated at between 3000 and 4000 million years old, are metamorphic. However not all metamorphic rocks are old, nor old rocks metamorphic. The process depends on the rocks being subject to heat and pressure on a large scale, and this usually happens near subduction zones. The fold mountain ranges which result when continents collide will contain large volumes of metamorphic rocks, as do the present day Himalayas and Alps, and here the metamorphic rocks are all usually less than 50 million years old. Wherever we find old metamorphic rocks on a large scale they usually indicate the presence of an old fold mountain chain. It may have been eroded away so that it is no longer as high as it was, but the structures in its deep root zone will still be there. In the north-western Highlands of Scotland large metamorphic areas are evidence of an ancient chain called the Caledonian Mountains which were formed around 400 Ma, and which extended into Greenland, Scandinavia and the north-eastern regions of North America.

All rocks, but particularly sedimentary rocks, can be affected by compression to form folds. Simple upfolds and downfolds are called anticlines and synclines (fig. 10), and more complicated fold structures can also be formed. A simple analogy is to push a table cloth across a table and against an obstruction and notice the various folds which are formed. Compression can also cause rocks to fracture and move along dislocation planes called faults. Compression will cause reverse faults and thrusts (fig. 10) whereas tension will cause normal faults as in rift valleys. Movement along fault planes can be vertical or horizontal, the latter movement occurring along what are called tear or wrench faults.

FIG. 10. *Folds and Faults.*

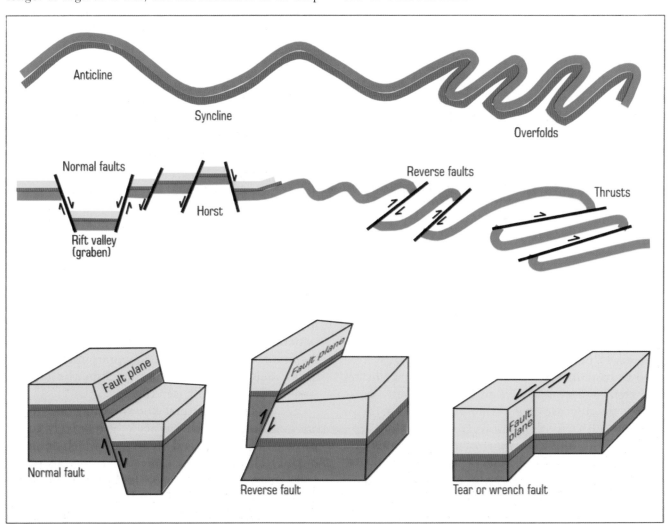

but everywhere there is enough original stratigraphy left to work out a detailed sequence. However, the effects of two later episodes of mountain building (orogenies) have formed enormous and complex fold structures (fig. 17) which can make it difficult to understand local sequences.

The early Dalradian is a relatively thinner sequence than that which follows later, and the Grampian and Appin Groups comprise 4 km of metamorphosed sedimentary rocks, sandstones, shales and important limestones – now schists and marbles. These rocks have been shown to follow on top of the Moine Schists around the Great Glen Fault and are not older than 850 Ma. They accumulated on a shallow shelf, with stromatolite (algal) limestones occurring on Islay. This shelf migrated north-west and reached the far north-west by the early Cambrian at 540 Ma.

The overlying Argyll Group contains at its base a remarkable sequence of beds indicating a late Precambrian glaciation – the famous Port Askaig Tillite (a tillite is the name we give to an ancient glacial deposit produced and then left behind by advancing and retreating ice sheets). The main deposit is a rock type called boulder clay – as its name suggests, a clay full of boulders. The clay is a rock 'flour' produced by the grinding action of the ice and the boulders it carries within its base. This particular

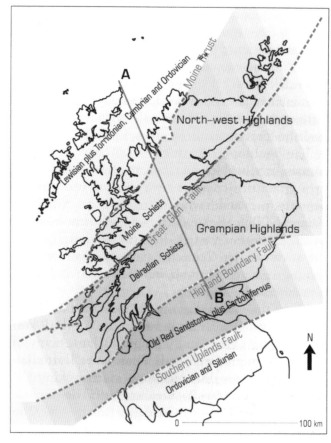

FIG. 16. *Geological terranes, major rock formations, and faults of Scotland. Fig. 17 shows cross-sections from A to B.*

FIG. 17. *Cross-sections through northern Scotland and the Grampian Highlands along line A–B on fig. 16. Fig. 17a shows the stratigraphy and radiometric dates. Fig. 17b shows the style and complexity of folding following 70 per cent crustal shortening in places during Grampian and Caledonian, and earlier orogenies. Fig. 17b after Open University Course S236, Block 6, Historical Geology p. 24, fig. 14. (Reproduced with permission of the McGraw-Hill Companies.)*

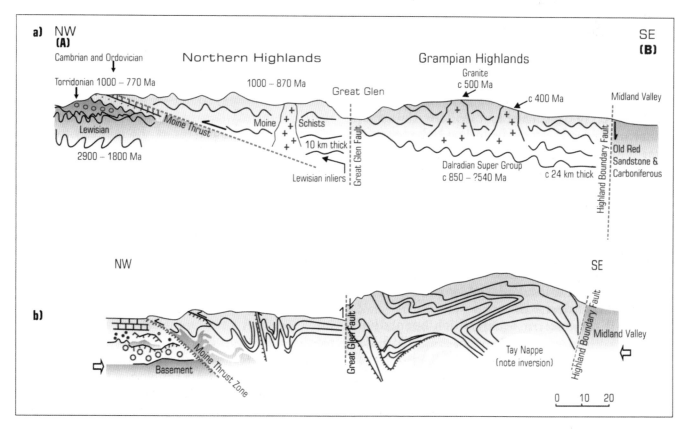

glacial sequence contains evidence for over forty separate ice advances. Even more remarkable is the magnetic evidence that this glaciation occurred within 10–15° of the equator, within a sequence of shallow water dolomites, calcium-magnesium carbonates. This Precambrian glaciation, dated around 600 Ma, may be unique in that it affected all latitudes; we have evidence for it from other parts of the world.

The boulder clays, thickest on Islay at 750 m, contain large boulders and appear to have been formed by a grounded ice sheet in a shallow sea. Similar deposits in Scandinavia, where they are called the Varanger Tillite, show evidence of movement from the south, as do the Scottish deposits. We imagine that the two areas were connected, but as we shall see, they became separated by a wide and deep ocean during the Cambrian and Ordovician.

Above the glacial deposits we find a further 20 km thickness of Dalradian rocks within the Argyll (9 km) and Southern Highland Groups (11 km), many of which are metamorphosed. These later Dalradian sediments are very thick and show evidence of initially being formed in a rapidly subsiding shelf. They include sandstones like the Jura Quartzite, which is 5000 m thick, as well as thin limestones. Later on the deposits show evidence of a much deeper ocean basin. A famous granite at Ben Vuirich intrudes the Argyll Group and is dated at 590 Ma, so we know the tillites are older than this. Recent work on the dolomites around the tillites has provided evidence for the earliest metazoan animals, possibly marine worms.

Within strata 8 km above the tillites we find a sequence of submarine basalt lavas, the Tayvallich Volcanics dated at 594 Ma. These lavas occur above the Tayvallich Limestone on the Argyll coast south of Oban, and may be associated with sea-floor spreading and a new ocean forming between Scotland and Scandinavia. This subsiding ocean basin then received a further 11 km of sediments in the later Dalradian Southern Highland Group. The highest beds contain a limestone, the Leny Limestone, which has Middle Cambrian trilobites in it, so we know that this late Dalradian episode of deposition in widening ocean basin went on into the Cambrian. The sediment for these basins could not have come from the north-west, for this was the even shallower shelf area in which the sandstones and dolomites of the far north-west accumulated near to the equator. Perhaps a land mass formed in the Midland Valley of Scotland area to provide the Dalradian sediments.

We call the ancient ocean which opened up between Scotland and Scandinavia at around 595 Ma the Iapetus Ocean. It spread to include all areas to the south towards the south pole areas and it is near here that southern Britain, England and Wales, could be found around this time. The story of the opening and closing of this ocean belongs to the Cambrian, Ordovician and Silurian Periods.

The Grampian orogeny

Orogenesis is the formation of fold mountains which is now explained in plate tectonic terms by subduction and continental collisions. Many of the Dalradian rocks were metamorphosed during an episode called the Grampian orogeny in the early Ordovician Period at around 500 Ma. They contain a variety of metamorphic rock types including marbles, slates and schists. Some of the already metamorphosed Moines were altered for a second time.

Metamorphic rocks are produced by changing existing rocks by heat and pressure. Modern laboratory experiments show that the temperatures involved are between 500 and 1000°C (above this rocks begin to melt), and the pressures involved are up to 20 000 atmospheres. This suggests that for metamorphism to occur rocks need to be deeply buried within the crust, probably at depths around 10–20 km. Burial can be achieved by the accumulation of great thicknesses in subsiding marine basins as described previously in the Dalradian. Metamorphism also occurs where sediments are dragged down into subduction zones; and above these zones due to the high temperatures and pressures which are produced.

The Dalradian sequences contain a variety of metamorphic rocks including rare marbles derived from limestones, quartzites and granulites derived from sandstones and a remarkable sequence of slates and schists derived from shales and mudstones. In the early 1900s Barrow studied the metamorphic rocks of the Grampians produced from shales, and worked out a series of metamorphic zones (called Barrow's zones) each with different minerals and indicating increasing grades of metamorphism (fig. 18).

When pressure and heat are first applied to a shale (a fine-grained laminated mud rock with clay minerals), the pressure causes the minerals to rotate and line up in one direction perpendicular to the stress. This causes the rock to split perfectly in only one direction called the cleavage, and the rock is now a slate (chapter 2, fig. 32). If the rock is subjected to more heat and pressure the minerals start to react with one another (over thousands of years) and new minerals are formed, particularly shiny green chlorite. The rock is now called a phyllite. With higher temperatures and pressures (and maybe the passage of water through the rock), other minerals appear including shiny white and black mica, and beautiful red garnets. The rock is now quite coarse grained and is called a garnet mica schist. At higher temperatures and pressure other minerals appear including blue kyanite. Many parts of the Highlands are made of mica schists with other minerals. Further metamorphism

two systems were ill defined and disputes arose. Murchison suggested that much of Sedgwick's Cambrian was in fact Silurian. The two fell out and rarely spoke after 1840, despite the fact that in 1839 Murchison had dedicated his epic work *The Silurian System* to Sedgwick with these words:

> To you, my dear Sedgwick, a large portion of whose life has been devoted to the arduous study of the older British rocks, I dedicate this work. Having explored with you many a tract, both at home and abroad, I beg you to accept this offering as a memorial of friendship, and of the high sense I entertain of the value of your labours.

Thus a great Victorian controversy arose, and as with the one that was later to affect Darwin, people fell into two camps. In the end (and after forty years!) Professor Charles Lapworth, himself an eminent geologist, in 1879 proposed a new system called the Ordovician, to cover the rocks between the Cambrian and Silurian which were in dispute.

This was not a compromise to save people's reputations, but a decision based on a scientific argument. Lapworth was able to show that the Ordovician contained a distinct set of fossils, different from the Cambrian and the Silurian. This is the fundamental definition of a geological system. The Ordovician essentially covered what Murchison had for some time been calling the Lower Silurian.

Cambrian plate tectonics and ancient geography

We can gain an insight into the position of the British area throughout geological time by looking at five important lines of geological evidence. First, there is the evidence for the ancient climate as indicated by rock types and fossils. For example desert sandstones, glacial boulder clays and fossilised coral reefs will tell us about past climates and probable past latitudes. This is done by comparison with modern climates, assuming that conditions have not changed a great deal in the last 2000 million years.

Secondly, fossils can tell us, through what Darwin called faunal and floral provinces, about ancient land bridges and land separation. Areas of land may be isolated from others by wide oceans or impassable deserts, which can prevent animal and plant groups from mixing and interbreeding, so that isolated communities arise. For instance the presence and survival of the Australian marsupial mammals suggests initial connections with other continents when marsupials first evolved, followed by isolation for a long period to protect them from the competition of the later more successful placental mammals.

Thirdly, palaeomagnetism (ancient magnetism)

allows us to establish where rocks were when they took up their initial magnetic field in relation to the magnetic poles. This tells us the ancient latitude of formation of many rocks. In fig. 26, which demonstrates the earth's magnetic field, three lavas formed at localities A, B and C will have iron-bearing minerals in them magnetised with directions as shown. This magnetism is then 'frozen' into the rock, and can only be destroyed or altered by heat. If continental drift occurs and A moves to B we can still see by its magnetic field that it originated near to the magnetic north pole. When measuring the magnetic orientation of A in its new position at B we have to consider its original physical orientation – whether it was vertical or horizontal, etc.

Fourthly, the presence of acid and intermediate volcanic lavas and ashes, rhyolites and andesites in a marine sedimentary sequence suggests that the rocks were formed near a subduction zone in a closing ocean. Conversely the absence of andesitic volcanic rocks suggests a widening ocean.

Fifthly, the presence of basaltic pillow lavas erupted on the ocean floor indicate old oceanic crust. Being dense, these are usually subducted at plate boundaries

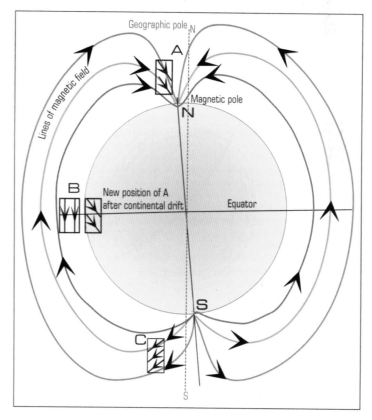

Fig. 26. *The use of palaeomagnetism in fixing ancient latitudes. Three lavas formed at A, B and C, have different magnetic field directions 'frozen' into them on crystallisation. If A formed near the north magnetic pole then moves, due to plate tectonics, to near the equator, the direction of its magnetic field will show it originated near to the pole and the distance it has subsequently travelled can be calculated. Magnetic reversals (fig. 3) will have an effect on this but a wide range of readings will show whether changes were in the northern or southern hemispheres.*

but sometimes appear within continents and are called ophiolites. When found in ancient rocks, these indicate that sea-floor spreading has happened in that area.

Using evidence from all of these five criteria geologists have built up an idea of the position of the British area during the Cambrian and succeeding periods. When discussing the Dalradian rocks of the

FIG. 27. *Cambrian palaeogeography. a: Late Cambrian (c 520 Ma) world geography showing widening Iapetus Ocean and ancient latitudes for Scotland (20°S) and England and Wales (60°S). During late Precambrian (600 Ma) Baltica was still attached to Laurentia. b: Cross-section through Iapetus Ocean during late Cambrian.*

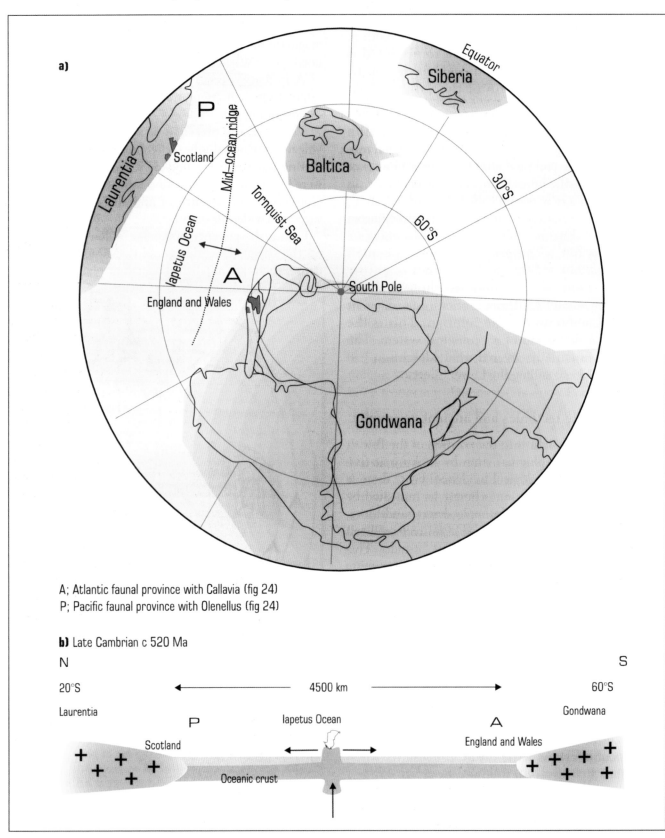

A; Atlantic faunal province with Callavia (fig 24)
P; Pacific faunal province with Olenellus (fig 24)

b) Late Cambrian c 520 Ma

(fig. 37) started to form during the later Ordovician as subduction continued on the Laurentian margin. By the end of the Ordovician, Laurentia and Avalonia-Baltica were perhaps only 1000 km apart.

Ordovician sequences in England and Wales

The British Ordovician, particularly in Snowdonia and the Lake District, is dominated by volcanic rocks, often thousands of metres thick. These volcanic rocks were erupted from ancient volcanic island arcs and marginal volcanic basins, and are inter-bedded with marine sediments often full of fossils.

North Wales

In North Wales volcanic rocks form the arcuate outcrop from Snowdonia around east of the Harlech Dome and towards Cadair Idris. This has been called by some authors the 'firey ring' of North Wales (chapter 2, fig. 29). The volcanic rocks occur at particular levels (horizons) within the seventy million years of the Ordovician Period and at other times marine sedimentary rocks separate the lavas and ashes. In a number of places volcanic rocks can also be seen to pass laterally into sediments. The volcanicity ranges from the Tremadoc Epoch up into the early Caradoc Epoch where it reached a climax in the Snowdon area.

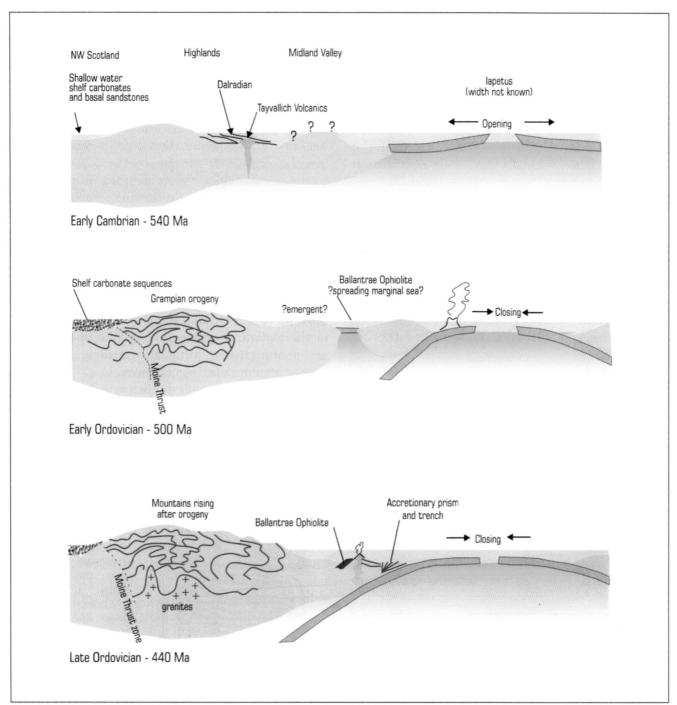

FIG. 37. *Close-up of boundary between Laurentia and the Iapetus Ocean during the Cambrian and Ordovician Periods. After Open University course S236, Historical Geology, Block 6, p. 29, fig. 19.*

The volcanicity died away at the end of the Caradoc Epoch and the latest Ashgill Epoch is represented only by sedimentary rocks – sandstones, shale, and thin limestones.

The start of the Ordovician Period in North Wales was marked by a marine episode carrying on from the Cambrian, indicated by thick muds, now metamorphosed to the Tremadoc Slates. The early Ordovician Tremadoc Slates have trilobites in them which have more in common with those in the later Ordovician than those from the underlying Cambrian. This is why the Tremadoc Series is now classified as Ordovician, whereas up to the 1960s it was considered Cambrian, according to Lapworth's original definition. The metamorphism of shales to slates (chapter 2, fig. 32) produces not only cleavage but also a geometric distortion in the rock by shearing. Thus a cube of shale can be deformed to a rhombohedron. A well known trilobite *Angelina sedgwickii* occurs in the slates but always has a deformed shape (fig. 38). No one knew what the original shape looked like until the same trilobite was discovered in the Shineton Shales of Shropshire. These are undeformed shales equivalent to the Tremadoc Slates, and a study of the undeformed specimens of *Angelina* tell us how much distortion the Tremadoc Slates have suffered.

A period of Earth movements then tilted the late Cambrian and early Tremadoc rocks so that a marked unconformity occurs at the base of the overlying Arenig Series. The sandstones with a basal conglomerate spread across the area, and on Anglesey rest on Monian rocks. The Cambrian and Tremadoc rocks on Anglesey have, as I say, been cut out by the unconformity.

The North Wales volcanic rocks have been widely studied. They contain island arc volcanic rocks and also lavas and ashes formed by volcanoes in marginal basins. Both subaerial and submarine lavas and ashes occur, but acid (rich in silica and hence explosive) ashes and ash flows predominate. Ancient calderas and vents have been found in Snowdonia and in many areas the ashes and lavas are intruded by sheets of igneous rock associated with the magma chambers, such as the famous 300 m thick granophyre sill (see below) which forms the north face of Cadair Idris (fig. 39). The fact that shallow-water fossils such as brachiopods are found in some of the ashes shows that many of the ashes fell into, or were erupted into, marine areas. Acid rhyolite and intermediate andesite lavas occur but it is now thought that many of the so-called lavas are the welded parts of ash flows called ignimbrites. These were laid down by high-temperature glowing cloud eruptions (nuées ardentes) which raced down the volcano flanks at high speeds. Some of these are associated with marine sediment and it is thought that some flows were actually formed as submarine emissions which flowed out onto the sea floor supported by their own high density. Pillow lavas occur, as do volcanic mud flows (lahars), and pyroclastic ash falls merge into ashy sediments with marine fossils. These marine deposits clearly flanked every emerging volcanic island.

In North Wales the volcanicity started in the Tremadoc Epoch around Rhobell Fawr, north of Cadair Idris, where 4000 m of subaerial basalt lavas are dated at c 508 Ma. The volcanicity continued into the succeeding Arenig, Llanvirn and Llandeilo Epochs around Arenig, Cadair Idris and the Aran Mountains. In the Arenig area volcanic ashes alternate with fossiliferous shales and sandstones, and around Cadair Idris the Aran Volcanic Group contains up to 2000 m of alternations of acid and basic lavas and ashes and sedimentary rocks, intruded by a thick sill of fine-grained granite called a granophyre.

Moving eastwards into the Berwyns the sequence is mainly Llandeilo-, Caradoc- and Ashgill-age sedimentary rocks with a limestone of Llandeilo age at the base. The sequence however contains up to six

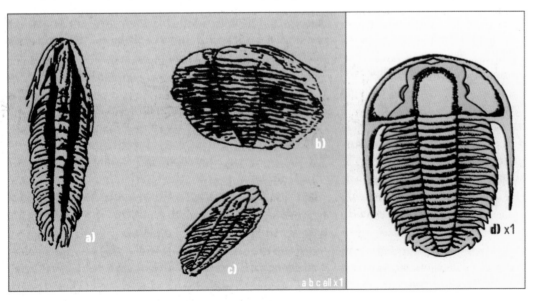

FIG. 38. *The trilobite Angelina sedgwickii from the Tremadoc Epoch. a, b and c: Deformed specimens from the Tremadoc Slates of North Wales. d: Undeformed specimen (reconstructed) from the Shineton Shales of west Shropshire. After Fortey & Owens, Geology Today, 1992, 8 pp. 218–221, fig. 2.*

famous Durness Limestone formed in the north-west, passing south into a deeper oceanic area. This oceanic area contained island arc volcanoes, but the exposures of volcanic rocks are now very limited. They are only found near the Highland Boundary Fault, on the Ayrshire coast at Ballantrae and in isolated outcrops in the Southern Uplands.

During the later Ordovician the rising Grampian Highlands caused by the Grampian orogeny cut off the carbonate platform from the southern areas, and from then on deposition in the south of Scotland was on a marine shelf and basin flanking the rising mountain chain. This marine basin in the south probably had an active volcanic island arc nearby, as many of the Ordovician greywackes contain mineral fragments derived from such volcanics, and these appear to have travelled from the south.

North-west Scotland
The Durness Limestone was described in chapter 2 as it ranges in age from early Cambrian to early Ordovician. There are 1300 m of dolomitic limestones containing stromatolite algal reefs, rare trilobites and brachiopods.

Along the Highland Boundary Fault are found isolated fault slices of Ordovician rocks with uncertain relationships to surrounding areas. Serpentinites and pillow lavas possibly represent a fragment of oceanic crust similar to that at Ballantrae. Black shales associated with the volcanics contain early Ordovician (Arenig–Llanvirn Epoch) fossils. There are also small thicknesses of late Ordovician sedimentary rocks.

Ayrshire–Girvan and Ballantrae
At Ballantrae the famous Ballantrae Igneous Complex contains a variety of igneous rocks – serpentinite, pillow lavas and associated cherts, which suggests it is part of an ophiolite, a rare fragment of basaltic oceanic crust which has been added to the continental margin during earth movements above a subduction zone. It is said to have been obducted in contrast to 99 per cent of dense oceanic crust which is subducted back into the mantle (fig. 37). Ancient ophiolites provide an opportunity to study a profile through a piece of oceanic crust on land. At Ballantrae the sequence has been deformed but there are good examples in the Alps and the Troodos Mountains in Cyprus.

The Ballantrae ophiolite could be oceanic crust, part of a marginal basin crust, or volcanic island arc material, but most opinions agree that it is oceanic crust of some sort. The complex, with its serpentinites and pillow lavas, also contains volcanic agglomerates, tuffs, cherts (banded silica) rich in fossil plankton called radiolaria, and black shales with Arenig–Llandeilo Epoch graptolites. Some of the pillow lavas are dated at 501–468 Ma.

Overlying the complex is a 2600 m thick classic sequence of fossiliferous sediments of Llandeilo, Caradoc and Ashgill age, which gradually encroached from south to north over the eroding igneous complex. This sequence was first studied by Charles Lapworth, who was able to show (Lapworth, 1878 and 1880) that it passes south and east into a much thinner black shale graptolite sequence in the Southern Uplands, where only 35 m of black shales at Dob's Linn equate to 3000 m at Girvan (fig. 53).

Basal conglomerates of the Barr Group are banked up against the volcanics, as are the overlying Llandeilo-age mudstones and limestones (Stinchar Limestone). The overlying Ardwell Group comprises mudstones and sandstones with Caradoc-age shelly fossils and a lateral equivalent limestone, the Craighead Limestone, with excellent fossils which rests directly on the Ballantrae Igneous Complex. Overlying Ashgill Epoch sediments complete the Girvan sequence. South of the Stinchar Valley towards the Southern Uplands Fault, the Ordovician sequence thickens to nearly 5000 m and contains mainly greywackes and conglomerates with shale partings with Llandeilo Epoch graptolites.

The Southern Uplands
Across the Southern Uplands Fault to the south and east of Girvan we come to a vast area of greywackes which stretches from the Rhinns of Galloway north-east to the Berwickshire coast. In this huge area the greywackes nearly always show very steep dips to the north-west, and in places show tight isoclinal folding (fig. 51 and chapter 4, fig. 71). During the late nineteenth century the British Geological Survey divided the area up into three large tracks of country trending north-east–south-west, called the northern, central and southern belts. The northern belt is mainly Ordovician, the central belt mainly early Silurian with Ordovician inliers, and the southern belt is late Silurian. There is thus an anomaly in that if the rocks all dip north-west, the youngest rocks should be in the north-west, but in fact the opposite is the case – with the oldest in the north-west and the youngest in the south-east! Graded bedding in the steeply dipping greywackes also indicates that the sequences young towards the north-west, and this adds to the apparent anomaly.

Work from the 1960s to the 1980s showed that in fact the greywacke belts are separated by high-angle reverse faults or thrusts (fig. 52) where younger rocks to the south have been thrust under older rocks to the north. On a larger scale the whole of the Southern Uplands is part of what is called in plate tectonic terms an accretionary prism. In this, wedges of sedimentary successive sequences on the descending plate above a subduction zone are pushed onto the

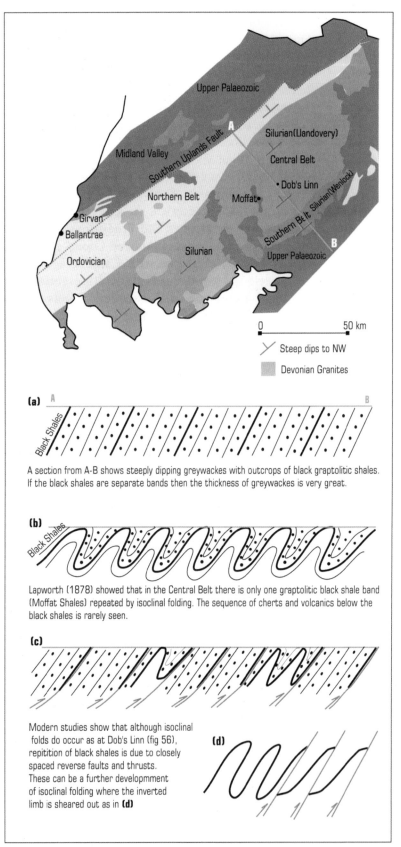

(a) A section from A-B shows steeply dipping greywackes with outcrops of black graptolitic shales. If the black shales are separate bands then the thickness of greywackes is very great.

(b) Lapworth (1878) showed that in the Central Belt there is only one graptolitic black shale band (Moffat Shales) repeated by isoclinal folding. The sequence of cherts and volcanics below the black shales is rarely seen.

(c) Modern studies show that although isoclinal folds do occur as at Dob's Linn (fig 56), repitition of black shales is due to closely spaced reverse faults and thrusts. These can be a further developmment of isoclinal folding where the inverted limb is sheared out as in **(d)**

FIG. 51. *Geological sketch map of the Southern Uplands and cross-sections illustrating various interpretations of the geological structures.*

continental margin with each wedge underthrusting the one in front (fig. 52). By this mechanism the oldest sequences are at the top of the pile, and the youngest at the bottom. This explains the anomaly. Within each individual faulted wedge the greywackes appear to young to the north-west, but the whole sequence youngs to the south-east. The accretionary prism theory was put forward in the 1970s but work in the late 1990s has suggested an alternative model with sediments formed on an extending continental margin rather than in the open ocean.

In the northern belt these wedges of steeply dipping greywackes contain, at the base of the sequences, pillow lavas and cherts together with black shales with Arenig-Llandeilo-age graptolites. These can be examined around Abington and Leadhills, where the pillow lavas are thought to be the same age as those at Ballantrae. Much of the northern belt contains unfossiliferous greywackes with thicknesses difficult to measure.

Recent evidence suggests that abundant detrital garnets in the làte Ordovician greywackes of the northern belt, and Highland Border Complex, were locally derived from Dalradian schists. This suggests that during the Ordovician, the Grampian, Midland Valley and Southern Uplands terranes were close together, as today. Large rivers flowed out of the rising Grampian terrane and across the Midland Valley into a Southern Uplands trench during the late Ordovician.

Further south, within the central belt, we find the classic localities of Dob's Linn, Hartfell and Glenkiln, all near Moffat, exposing a very thin sequence of black shales within what appears to be an enormous thickness of Silurian greywackes. It was around here that Lapworth showed that a black shale Ordovician sequence (Hartfell and Glenkiln Shales) which is only 35 m thick, with thin bentonites, contains graptolite zones which equate with 3000 m of shallow-water sediments at Girvan. This is what geologists call a condensed sequence (fig. 53). The black shales were formed in marine areas of poor circulation with no oxygen (anoxic). Graptolites (fig. 54) living as macro-plankton in the surface waters fell down on death in their thousands onto the black ooze on the ocean floor. Thus these black shales are

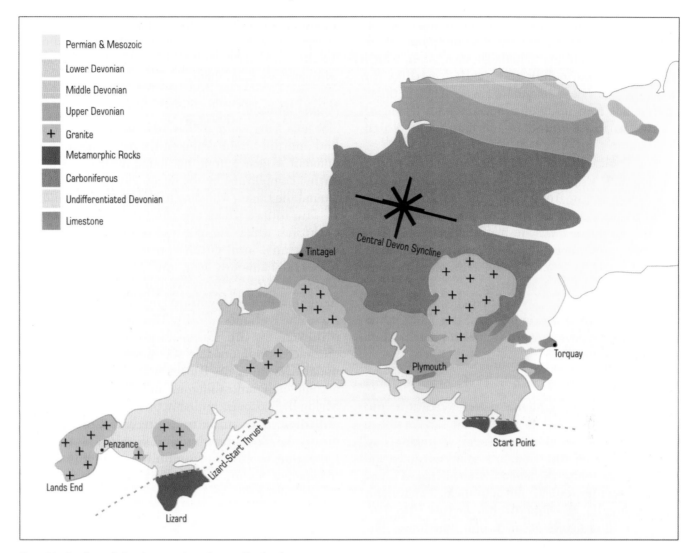

FIG. 78. *Geological sketch map of south-west England.*

South-west England

Devonian rocks outcrop in two areas (fig. 78) north and south of the central Devon syncline which along its axis contains Carboniferous rocks. The two areas show very different sequences, and in particular the northern areas show the appearance of Old Red Sandstone facies at certain horizons, which reflects the proximity of the northern landmass, and the frequent north–south migrations of the shoreline and coastal areas.

All the Devonian (and Carboniferous) rocks of south-west England have been affected and deformed by a major orogeny at the end of the Carboniferous. This also affected large areas of western and central Europe, and is given a variety of names – Armorican, Hercynian and Variscan. Variscan is the term used most commonly in Britain. During this orogeny many of the finer sediments of the south-west Devonian were deformed and metamorphosed to slates. Higher grades of metamorphism in the Lizard and Start Point areas are associated with major thrust planes which have brought northwards highly altered Devonian and older sediments, and volcanic and igneous rocks with oceanic affinities.

The famous granites of south-west England (fig. 78) intruded the Devonian sediments in the late Carboniferous and caused local thermal (heat only) metamorphism and mineralisation. The famous tin mines of Cornwall work veins associated with the granites and surrounding Devonian rocks. Much of this deformation and igneous activity will be discussed later (chapter 8).

South Devon
A well-displayed sequence of Devonian rocks can be studied in South Devon in a belt of country from Plymouth to Torquay. This sequence can also be traced westwards into Cornwall, where the structure is more complicated and the rocks less fossiliferous.

The oldest Devonian rocks are the Dartmouth Slates which represent an Old Red Sandstone coastal plain environment – the most southerly advance of this continental facies. At least 700 m of dark sandstone and slates have vertebrate and plant remains, and basaltic tuffs also occur. The succeeding Meadfoot Beds and Staddon Grits comprise 450 m of true marine shallow-water sandstones, slates and limestones, together with volcanic lavas and ashes. In the Middle Devonian the famous limestones of the Torquay, Plymouth and Chudleigh areas occur, which have been used widely as so-called marbles for fireplaces, etc. These limestones are up to 800 m thick, and include reefs made of stromatoporoids (sponges) similar to those found in Germany and North America.

Thick volcanic lavas and ashes also occur within the limestone sequence. The sponge reefs are followed by coral-bearing limestones and the carbonate sedimentation continued into the late Devonian. The highest Devonian rocks are marine mudstones, with goniatites, up to several hundred metres thick, except around Chudleigh where they are remarkably thin and were deposited over a marine rise.

In the far south of Devon, around Start Point and Bolt Head, south of Salcombe, the Start Thrust brings northwards a sequence of schists probably derived from Devonian sediments.

North Cornwall

The South Devon sequence passes westwards into Cornwall to form a belt of country from Looe across to the northern Cornish coast around Watergate Bay. However here all the sequences comprise slates and volcanics, including the famous Delabole Slates. These contain well-preserved, but deformed, brachiopods known locally as Delabole butterflies (fig. 79). Undeformed examples occur in thin limestones. The Devonian rocks here contain Middle and Upper Devonian fossils and appear to pass up without a break into Carboniferous rocks. The Devonian slates are strongly deformed and tight folds can be seen in the splendid coastal sections from Newquay northwards to Tintagel, where the sequence becomes Carboniferous in age. Around Pentire Point near Padstow 500 m of basaltic sub-marine pillow lavas occur interbedded with late Devonian slates.

South Cornwall

The area lying south of Newquay and including Land's End and the Lizard (Falmouth) area contains a very different sequence from that found further north. It is further complicated by the major east–west thrust belt around the Lizard, which continues east to Start Point.

The oldest rocks are the Ordovician Veryan Quartzites, which are of a type similar to those found in Brittany, and which can be seen around the Roseland–Falmouth area, and south of Mevagissey. These occur as thrust wedges in the Devonian sequence but are overlain unconformably by the Roseland Volcanics of early Devonian age. These are followed by the Gramscatho Beds which are greywackes with a middle 400 m unit of limestones of middle Devonian age. Further west towards Land's End a completely unfossiliferous unit of slates called the Mylor Beds occurs, probably of early Devonian age. All these rocks are intruded by late Carboniferous granites. The Lizard complex includes an early Devonian (c 400 Ma) ophiolite, a highly altered fragment of oceanic crust, welded onto the continental margin during the late Devonian (see chapter 8).

North Devon and West Somerset

North of the central Devon syncline occupied by Carboniferous rocks we find splendid coastal sections of Devonian rocks in North Devon and West Somerset. Up to 5000 m of Devonian rocks occur which show two major incursions (from the north) of non-marine Old Red Sandstone facies in an otherwise marine sequence (fig. 75). Although the main succession is clear, structural complexity makes it difficult to calculate accurate thicknesses. Marine goniatites are rare and correlation is done on other fossils.

The sequence starts off marine with the Lynton Beds, comprising 500 m of shallow-water slates and sandstones with marine shells. The overlying 1500 m of the Hangman Grits shows a continental development of braided streams and

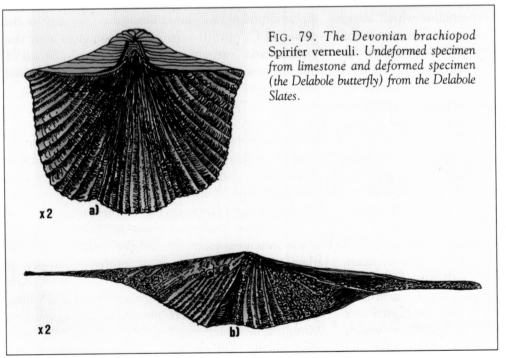

FIG. 79. *The Devonian brachiopod Spirifer verneuli. Undeformed specimen from limestone and deformed specimen (the Delabole butterfly) from the Delabole Slates.*

x2 a)

x2 b)

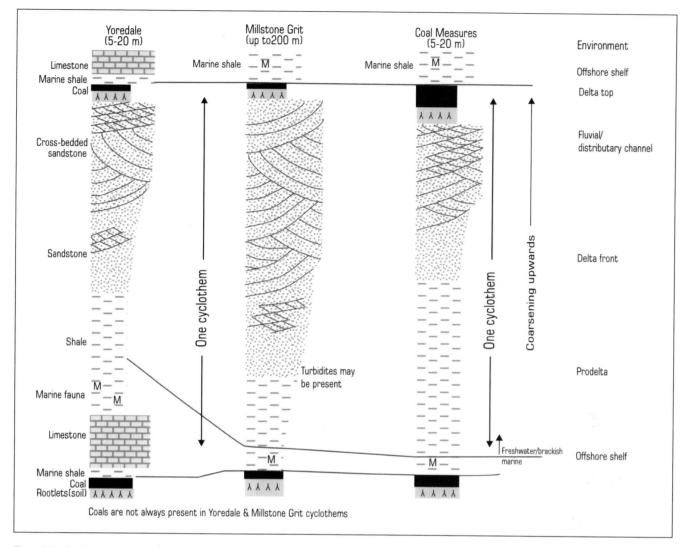

FIG. 87. *Carboniferous cyclothems. After Open University, Course S236, Block 6, Historical Geology, p. 41, fig. 28.*

but liable to subsidence and marine flooding. The delta systems simply move elsewhere into a new channel. It is probably that an explanation of Carboniferous cyclothems will include all of these three possibilities.

Plate tectonics and palaeogeography

The formation and widening of the Rheic Ocean in the early Ordovician eventually caused the closure of the Iapetus Ocean by the end of the Silurian. The subsequent closure of the Rheic Ocean (fig. 88) during the Devonian and Carboniferous Periods was a complicated process with a number of phases – late Devonian (Bretonic), mid-Carboniferous (Sudetic) and late Carboniferous (Asturic).

The Variscan orogeny is the name given to all these events, which eventually resulted in the closure of the Rheic Ocean and the formation of major fold structures, thrusts and granite emplacement over south-west England and many parts of Europe. Although the effects are most noticeable in Devon

and Cornwall, the orogeny caused a variety of folds and faults over southern Britain.

The Rheic Ocean during the early Devonian (fig. 88) separated the Old Red Sandstone Continent, which included North America, Greenland, Britain and Scandinavia, from the southern continent of Gondwanaland and what can be called proto-Asia. Fragments of micro-continents and marginal basins were to develop as western Europe within the Rheic Ocean. The growth of the huge proto-Pacific Ocean caused the closure of the Rheic Ocean during the Devonian and Carboniferous, and the collision of the northern and southern continental masses plus 'Asia' caused the formation of one supercontinent which Alfred Wegener called Pangaea (fig. 88). He called the large proto-Pacific Panthalassa. Where the two continental areas joined or sutured a large fold mountain range was formed, from the Appalachians through into south-west Britain, central Europe and the Ural Mountains (chapter 8, fig. 101). A large area of north-west Africa was also affected. The Ural Mountains are the natural geological boundary in Russia between Europe and Asia. In Europe this fold mountain range is called the Hercynides, and its formation at the end of the Carboniferous is similar to

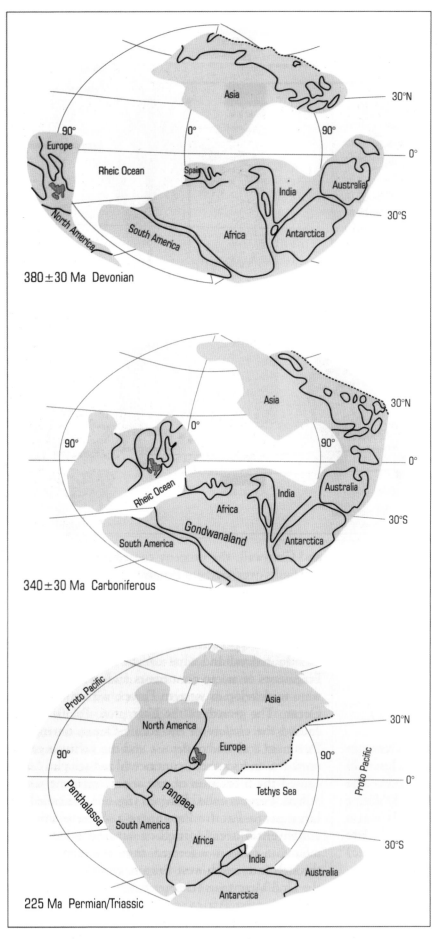

380±30 Ma Devonian

340±30 Ma Carboniferous

225 Ma Permian/Triassic

FIG. 88. *World palaeogeography for the Devonian, Carboniferous and Permian, showing closure of the Rheic Ocean. After, Open University, Course S236, Block 6, Historical Geology, p. 8, fig. 3.*

that of the Caledonides at the end of the Silurian; it was caused by ocean closure and continental collision. By the end of the Carboniferous, Britain lay within the arid heart of Pangaea.

Stratigraphical framework

The Carboniferous Period is divided into four epochs and the Carboniferous System therefore into four equivalent series as shown in fig. 89. Of these the Dinantian Series roughly equates to the Carboniferous Limestone, the Namurian to the Millstone Grit and the Westphalian to the Coal Measures. It is unlikely that Stephanian rocks occur in Britain. The Dinantian Carboniferous Limestone contains a well-known coral and brachiopod assemblage and this was used by Vaughan in 1905 to divide up the marine Carboniferous Limestone into a classic zonal sequence in the Avon Gorge area of Bristol, where the limestones are well exposed (fig. 90). However corals and brachiopods are very much dependent on bottom conditions and depths of water and so the zones are difficult to apply in some deep-water areas. Nevertheless the scheme has been widely used.

A more novel classification was set up in the 1970s, which produced six stages within the Dinantian based on cycles of sedimentation (fig. 89). These cycles were brought about by changes in sea level and each has a marine transgression (advance of the sea) at its base and a regression (retreat) at its top. These six cycles are called mesothems and are best studied at the margin of basins. Within each mesothem there can be numerous smaller cycles of sedimentation.

Similar cycles have been identified in the Namurian and used to identify stages (fig. 89), but here marine goniatites are useful as well. Sixty marine bands have been identified in the Namurian and the zones recognised all over the northern hemisphere. Sixty zones in 12 million years (the time scale of Namurian) equates to a new goniatite zone every

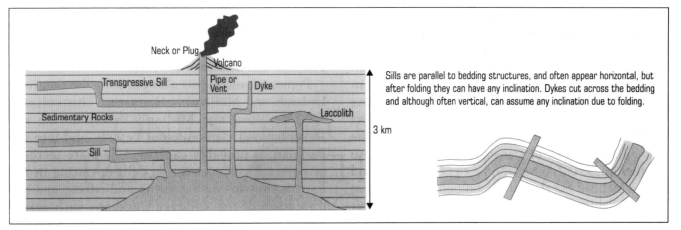

FIG. 100. *Formation of dykes and sills. The magma is usually basaltic, a mobile (runny) type which can travel great distances, and hardens to form basalt or dolerite.*

Dinantian is represented by the Cementstones which are impure limestones of a brackish, sometimes very saline, lagoonal type. These follow on from the highest Old Red Sandstone rocks and the Kelso Lavas. The Cementstones are followed by sandstones laid down by braided rivers, the Fell Sandstone. The succeeding Scremerston Coal Group contains the earliest coals in Europe. The Dinantian sequence is completed by typical Yoredale type sediments including the Dun Limestone which can be traced southwards onto the Alston block, where it equates with the Melmerby Scar Limestone. Further west along the northern Solway Firth Dinantian rocks include very coarse breccias derived from local granites, as well as limestones.

In the Midland Valley the basal Carboniferous Calciferous Sandstone Group includes mainly cementstones in the west of the area, which pass east into 1500 m of deltaic sandstones. Large scale eruptions occurred during the early Carboniferous to form the Clyde Plateau Basalts which are up to 1000 m thick. These formed a barrier between the western area and an eastern basin where in the Lothians organic rich Oil Shales were deposited. Typical Oil Shales comprise around Edinburgh 1800 m of sandstones and shales with the true oil shales made up of thinly laminated bituminous muds. In the 1850s James 'Paraffin' Young developed an extraction process to produce crude oil from the oil shale seams. This left behind spoil heaps of red shales called bings, which can clearly be seen west of Edinburgh.

In the Midland Valley the Namurian rocks comprise the Lower and Upper Limestone Groups with the Limestone Coal Group, up to 500 m thick, in between. This sequence is similar to the Yoredales of northern England but in the Limestone Coal Group limestones are thinner and the upper part of the cyclothems with shales, sandstones and coals are better developed with workable coal seams. Volcanic ashes are common at various levels within the Namurian and the so-called Passage Beds, with fluvial sandstones, marine mudstones and basalt lavas in Ayrshire, pass up into the Westphalian Coal Measures.

The Westphalian Coal Measures of Scotland resemble those of England and Wales and are found in the old coalfield areas of the Midland Valley, as well as in coalfields across the Southern Uplands Fault, on Arran and in parts of the south-west Highlands area. In Ayrshire the productive Coal Measures are 700 m thick and pass up into red sandstones and shales, indicating arid conditions at the end of the Carboniferous.

Large areas of the Midland Valley are occupied by Carboniferous lavas and ashes and there are numerous igneous intrusions, dykes and sills, and old volcanic vents scattered about the area.

Molten rock or magma is often contained within magma chambers at fairly shallow depths within the Earth's crust, 2–5 km down. These chambers feed magma upwards through volcanic vents to volcanoes, where it erupts onto the surface. However magma does not always reach the surface. It can move up from magma chambers through vertical fissures and harden within the surface layers of the crust to form vertical sheets of igneous rock called dykes (fig. 100). These can vary in thickness but are usually a few metres across, and because the magma has to be fairly mobile to move along fissures the dyke is often a basalt.

Dykes can occur as swarms, and many hundreds are found in Scotland cutting through the sediments and metamorphic rocks of the Highlands. Swarms include many dykes all trending in one direction. On the south coast of the Isle of Arran numerous dykes of Tertiary age cut the older rocks, often ten or so in 100 m. Dykes are also common in the Midland Valley Carboniferous rocks. In oceanic areas swarms of dykes feed basalt to the volcanic mid-ocean ridge crests, and can be examined where oceanic crust is preserved on land as at Ballantrae in Ayrshire.

Magma can also travel horizontally along bedding planes, particularly near to the earth's surface, and horizontal sheets of igneous rock within sedimentary rocks are called sills (fig. 100). The mobile magma actually forces the bedding apart, and also bakes the rocks on either side of the intrusion.

James Hutton proposed that the igneous rock on Salisbury Crags (fig. 99) below Arthur's Seat in Edinburgh was intruded into the surrounding sedimentary rocks and suggested that the baking effect around the margins proved the intrusive nature of the sill. Sills can move or transgress from one level to another (fig. 100). They can be tipped up later and appear inclined but the basic definition of a sill is that it is parallel to the bedding, whereas a dyke cuts across the bedding.

The vent of a volcano is usually filled up with volcanic ash and hardened lava when the volcano becomes extinct. These necks or plugs often survive erosion when the surrounding areas are worn away. They form upstanding circular crags, and many necks and plugs are found in the Midland Valley as remnants of the Carboniferous volcanic activity. The most famous is the one on which Edinburgh Castle is built.

The Clyde Plateau Basalts are up to 1000 m thick and occur in the western areas of the Midland Valley in the later Dinantian Epoch. The volcano of Arthur's Seat in Edinburgh also appears to be of this age. In the northern part of the Clyde Plateau two prominent lines of volcanic necks occur, which presumably fed the basalt lava fields, one of them within the lavas of the Campsie Fells. In Edinburgh the castle is built on an old volcanic plug and at Arthur's Seat volcanic necks with coarse infilling debris (agglomerate) cut through volcanic lavas. The nearby Salisbury Crags are formed by a sill of basalt. This is a sheet of igneous material intruded parallel to the bedding of a sedimentary rock which has forced the bedding planes apart (fig. 100). A number of volcanic necks and plugs occur around the Firth of Forth, including the Bass Rock and North Berwick Law (chapter 13, fig. 170). Numerous volcanic necks associated with Dinantian lavas occur in the Borders between Dumfries and Kelso.

Namurian volcanic ashes occur in the Midland Valley and Westphalian lavas are found particularly in the east.

Numerous dykes (3000 or more) occur in the south-western Highlands and are probably of late Carboniferous or early Permian age. There are also numerous late Carboniferous to early Permian dykes and sills in the Midland Valley. One particular sill associated with dykes is called the Midland Valley Sill Complex and outcrops around the Firth of Forth extending over (and underlying) an area of 1600 km². It is a dolerite, which is really a medium-grained basaltic rock, and the sill is probably the same age as the famous Whin Sill of northern England which intrudes early Carboniferous rocks in the northern Pennines and is dated at c 300 Ma.

Permian basalt lavas also occur in the Southern Uplands and Midland Valley areas. This volcanic activity along the Midland Valley during the Devonian and Carboniferous Periods, and into the Permian, is probably explained by plate tectonic processes. Deep convection cells operating in the Earth's mantle cause doming and rifting and the formation of rift valleys or grabens with active volcanoes. These can further widen to form new ocean floors as in the modern Red Sea, but others just remain as rift valleys with volcanic activity but never widen. This is true of the Rhine rift valley and, as we shall see later, an important rift under the North Sea which formed in the Jurassic Period. The Midland Valley during the Devonian and Carboniferous probably had similarities with the famous East African Rift Valley where volcanic activity is widespread but no one is sure whether it will widen to an ocean or not.

Summary

The Carboniferous Period, 363–290 Ma, saw an initial spreading of a warm, shallow tropical sea over the eroded remnants of the Caledonian Mountains and the Old Red Sandstone Continent. This sea deposited the Carboniferous Limestone over many parts of Britain, but in the northern Pennines limestones are part of cyclical sequences which also contain clastic sediments caused by rapid sea-level changes. In south-west England deeper Rheic Ocean sediments without limestones form the Carboniferous Culm Measures. In Scotland the early Carboniferous lagoonal limestones and fluvial sandstones also contain great thicknesses of basaltic lavas, and this volcanicity continued throughout the Carboniferous.

The early Carboniferous seas were invaded by rivers and deltas forming the Millstone Grit during the middle of the period, and during the later Carboniferous these supported luxuriant swamps and rain forests. The deltas and their swamps were repeatedly covered by marine and lagoonal sediments caused by rapid changes in sea level. The buried peat deposits quickly decayed to form coal seams within sandstones and shales of the Coal Measures.

Towards the end of the period the Variscan orogeny (see chapter 8) caused the British areas to rise above sea level and a change from a humid to an arid climate caused the formation of continental sediments.

CHAPTER 8

ONE SUPERCONTINENT, PANGAEA – THE VARISCAN OROGENY

The closure of the Rheic Ocean at the end of the Carboniferous had profound effects in Britain, but an even wider impact globally. It formed fold mountain ranges along the suture line (chapter 7, fig. 88 and fig. 101), including the Appalachians in North America, various mountains in north-west Africa and central Europe, the Pyrenees, and the Urals. However, none formed in Britain. The disappearance of the ocean caused by the widening of the proto-Pacific (chapter 7, fig. 88) meant that by the end of the Carboniferous only one supercontinent existed, and one super-ocean. Alfred Wegener called these Pangaea and Panthalassa (all land and all sea). The modern configuration of the continents and oceans is the result of the subsequent break-up of Pangaea from Triassic times onwards, and the movement of continents around the globe caused by the appearance of three new oceans, the Atlantic, Indian and Southern oceans.

In Europe the Variscan orogeny had definite phases, the last being late Carboniferous/early Permian. Some people refer to the Carbo-Permian orogeny and elsewhere in Europe it is called the Hercynian or Armorican, but Variscan is the title most used in Britain.

Nearer to home the Variscan orogeny caused intense folding and deformation, with slate grade metamorphism, over south-west England. However, there were no andesite lavas associated with subduction zones during the closure phase of the Rheic Ocean in south-west England, and no fold mountains formed, in contrast to the Caledonian orogeny. This has led to the suggestion that the ocean closed mainly as a result of lateral fault movements rather than head-on continental collision and subduction. In particular

FIG. 101. *Variscan fold belts. After, Open University Course S236, Block 6, Historical Geology, p. 7, fig. 2.*

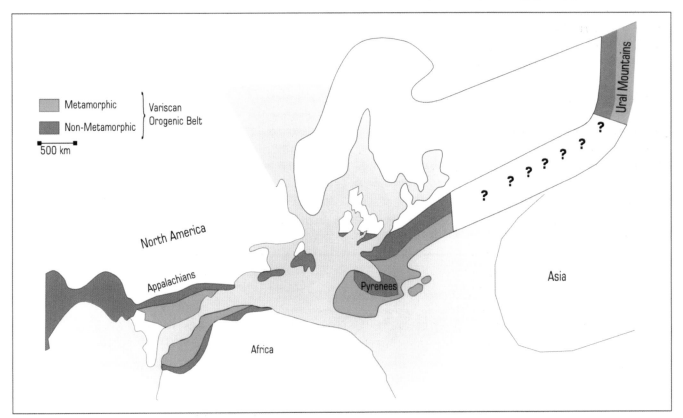

North America had to move laterally a considerable distance to make contact with north-west Africa (chapter 7, fig. 88). Deformation in south-west England was followed by the intrusion of a granite batholith dated at 270–280 Ma, i.e. early Permian. The granite cuts already deformed late Carboniferous Westphalian strata so we know the orogeny is post Westphalian/pre Permian.

To a lesser extent, the orogeny also affected South Wales, the Mendips and the Forest of Dean, where rocks are unmetamorphosed but can be intensely folded and affected by thrusts, as in Pembrokeshire. To the north of St George's Land the effects were even less noticeable but the coalfield areas are affected by Variscan faulting, and by early and late events within the Variscan orogeny termed Sudetic and Asturian. In the northern Pennines the blocks and basins were separated by faults that produced 'half-graben' effects (chapter 7, fig. 93). These faults were present at the start of the Carboniferous but were active during the Variscan orogeny. Large displacements on the western parts of this fault system, called the Outer Pennine Faults (chapter 3, fig. 50), now separate these areas from the younger rocks of the Vale of Eden. Faulting continued along the Midland Valley rift in Scotland during the Variscan but the volcanicity here was probably caused by doming, followed by rifting, over a mantle 'hot spot' or plume which was active from the Devonian onwards.

By the end of the Carboniferous the Variscan orogeny had caused uplift over the whole British area, which then lay within the arid hinterland of Wegener's Pangaea, but there were no huge mountains. The erosion of the Variscan highlands would provide the sediment for the rocks of the next geological period, the Permian.

South-west England

The main structure in this area is the east–west central Devon syncline with Carboniferous rocks in its core (chapter 6, fig. 78). Superimposed on this major fold are complex smaller fold structures and so the whole is called a synclinorium. Major thrusts affect the south of the area. The Devonian and Carboniferous rocks have been intensely deformed and folded into complex structures clearly visible in the coastal sections. Shales have been folded and metamorphosed to slates and then often refolded so that the cleavage is further deformed. There appear to have been two main episodes of deformation – early Carboniferous (Sudetic) movements dated at 340–320 Ma and late Carboniferous (Asturic) movements dated at 300–280 Ma. In south Cornwall a late Devonian (Bretonic) event can be dated at 365–345 Ma. Major compression appears to have come from the south, but lateral movements probably took place as well. The Lizard Thrust has brought up from the south a complex of ultrabasic igneous rocks, pillow lavas and serpentines; which could in part be an ophiolite, a fragment of oceanic crust, together with gabbros, gneisses and schists. However, although the age of the complex is in doubt, recent work has dated the Lizard ophiolite as early Devonian, at around 400 Ma, and suggests it was emplaced (obducted onto the continental margin) in the late Devonian Period, as part of the early Variscan orogeny. Further east along similar thrusts the Dodman Phyllites and Start Schists are faulted against Devonian rocks (chapter 6, fig. 78).

The Variscan events in south-west England were followed by the intrusion of a large granite batholith dated at 280–270 Ma. Erosion has exposed surface expressions of it to form the famous granite areas of Dartmoor, Land's End, Bodmin Moor, etc. Underneath these areas the granite is continuous and joins up with the other outcrops including the Scilly Isles. It is a peculiarly shaped batholith (fig. 102) probably injected northwards as a thin sheet rather than from the melting of deep crustal layers.

Associated with the granite are two well-known but different types of mineralisation. Mineral veins rich in tin and copper can be seen to radiate outwards from the Land's End and Carn Manellis granites in particular. The once thriving Cornish tin industry exploited these veins at depth. Some of the veins appear to be late cooling effects of the granite, whereby solutions concentrated in metals and dissolved in water at high temperatures and pressures

FIG. 102. *Intrusion of south-west England granite batholith, and generalised Variscan structure. After, Open University, Course S236, Block 6, Historical Geology, p. 47, fig. 32.*

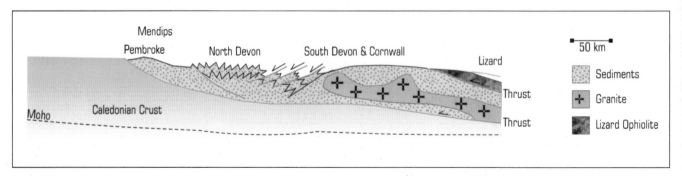

200 m of shales and sandstones, the Staffin Shales, with good ammonite faunas ranging up into the Kimmeridge Stage.

On the Sutherland coast of the Moray Firth Liassic rocks are found at Golspie, overlying the Rhaetic. Middle Jurassic rocks are found at Brora, where the famous Brora Coal occurs within sandstones and shales. Between Brora and Helmsdale we find excellent exposures of Upper Jurassic strata. 100 m of shales and sandstones are followed by the celebrated 700 m of Kimmeridge Clay, with bituminous marine shales with ammonites. Within the clays are remarkable boulder beds with huge blocks of Old Red Sandstone, Helmsdale Granite, gneisses and eroded Jurassic sandstones. These boulder beds formed next to a rising fault escarpment on the edge of the North Sea basin. The basin subsided along the Helmsdale Fault, which is a branch of the Great Glen Fault. We thus have evidence of late Jurassic fault movement on the edge of the North Sea basin in which thick Kimmeridge Clay deposits with organic material provided a good deal of the oil found today in sandstone reservoirs.

Summary

The Jurassic Period saw the onset of tectonic activity in the North Sea with the establishment of a rift valley, triple junction and associated volcanism. The Tethys Ocean spread a shallow epicontinental sea over most of Britain which laid down clays and thin limestones, the Lias, at the beginning of the period. Important chamosite and siderite ironstones formed in some areas. The early Jurassic seas were full of ammonites and marine reptiles.

Mid-Jurassic uplift produced a positive area in the central North Sea which fed deltaic and fluviatile sediments, the Estuarine Series, into the north of England, Scotland, and the north-east Midlands during the middle of the period. Meanwhile, the south of England was covered by shallow-water lagoons starved of land-derived sediment and extensive carbonates developed in the form of oolites, forming the Inferior and Great Oolites. Open marine areas only existed at this time in south-west England. Ammonites were abundant in the Lias but not so common in the oolites, and absent in the deltaic sediments of the north. Mid-Jurassic earth movements led to an unconformity within the Inferior Oolite, and local areas (axes) of uplift affected oolite sequences and thicknesses during the Lower and Middle Jurassic. A marine transgression produced a widespread limestone, the Cornbrash, at the end of the Middle Jurassic.

The Upper Jurassic saw the establishment of open marine conditions over all of Britain and thick organic-rich clays, the Oxford and Kimmeridge Clays, were laid down, separated by Corallian limestones. The Kimmeridge Clay is an important source rock for North Sea oil. The Upper Jurassic seas were full of ammonites, fish, ichthyosaurs and plesiosaurs. The collapse of the North Sea uplifted area and associated graben led to great thicknesses of Upper Jurassic coarse sediments in fault bounded basins in the North Sea.

At the end of the Jurassic, limestones, sometimes evaporitic, were restricted to southern England due to uplift in northern areas and a eustatic fall in sea level, but marine conditions continued into the Cretaceous in the Yorkshire Basin.

FIG. 132. *Chalk cliffs at the Seven Sisters, Sussex.* (Copyright, Landform Slides.)

CHAPTER 11

CHALK SEAS AND THE END OF THE DINOSAURS – THE CRETACEOUS PERIOD

The Cretaceous Period, named after the Latin for chalk, lasted for 80 million years, from 146 to 65 Ma, and is thus the longest of all the geological periods since the end of the Precambrian. It was set up in the nineteenth century on a sequence of rocks (a system) with diagnostic fossils, with no reference to absolute time – Victorian geologists would have had no idea as to how long the period lasted in terms of millions of years. It is an interesting thought that geological periods and systems whose boundaries are based entirely on fossil evidence, major extinctions, and changes in flora and fauna, average between 31 and 80 million years in length. Major changes in animals and plants have not happened in a random way but according to a reasonably regular pattern, and there are those who argue that major asteroid collisions with the Earth happen every ten million years or so.

Over southern and eastern England the Chalk (fig. 132) symbolises the Cretaceous System, as well as giving its name to it. The white cliffs of Dover and the rolling Downs are formed by one of Britain's most remarkable rocks – a pure limestone up to 500 m thick, and with no modern equivalent forming anywhere in the world. So here James Hutton's principle of uniformitarianism – that the present is the key to the past – breaks down. The Chalk stretches from Northern Ireland across Britain to Denmark and south into France, and also into central Europe. It is 98 per cent pure calcium carbonate and made up of countless million of tiny microscopic calcite plates of marine algae called coccospheres. Individual rings of plates are called coccoliths (fig. 133).

It was only when the electron microscope was invented in the 1960s that the structure of Chalk was identified in detail. It is a late Cretaceous formation, which formed as a calcareous ooze on the sea bed following algal blooms. Typical white Chalk is neither a deep oceanic ooze nor a deposit of shallow water. It was probably deposited under normal marine salinity at a depth of between 100 and 600 m. In the clear water of the modern tropics coccolithophorids live at depths of 50–200 m. There is little evidence of strong

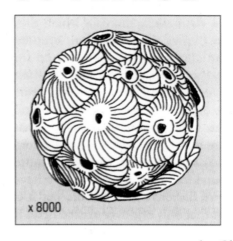

Fig. 133. A marine coccosphere, a type of algae. The Chalk is made up of countless millions of the individual minute circular plates called coccoliths.

scouring or current action in the Chalk and deepish water is suggested. However the presence of hexactinellid sponges suggests depths of 80–100 m. Modern calculation on the Chalk sequences in Britain show that between 20 and 40 m accumulated in a million years. That equates to a sedimentation rate of 1 cm in 500–1000 years.

The Chalk also contains abundant nodules of flint, which often occur as bands and can sometimes be seen replacing fossils. This process is called petrifaction and sea urchins and other fossils completely turned to flint can be found in the Chalk. A good source of petrified fossils are fields on the Chalk which have just been ploughed and which are covered with flints. It should be noted here that the so-called petrifying wells in Derbyshire use waters rich in calcium carbonate from the older Carboniferous Limestone which simply coats objects with limestone but does not petrify them.

Flints are formed of pure silica (silicon dioxide), better known as quartz. This is chemically very different from the calcium carbonate of the Chalk limestone, and for centuries people have wondered where the flints originated. The most plausible explanation is that they are concentrations of silica derived from siliceous sponges that grew on the sea floor, and other siliceous micro-fossils. After death the silica dissolved in sea water, was dispersed initially within the soft chalk sediments, and then collected around areas of organic concentration to precipitate out as nodules which replaced the Chalk sediment.

Early Cretaceous rocks include non-marine deltaic sediments in which dinosaurs are found. The name dinosaur ('terrible lizard') was first used by Sir Richard Owen, the founder of the Natural History Museum in London, in 1841, based on reptile remains from the Lower Cretaceous rocks of Sussex and the Isle of Wight. One of the earliest dinosaurs described was called *Iguanodon* (fig. 134). One of the newest was discovered in the Lower Cretaceous of Surrey, in 1983. An amateur geologist called John Walker, found a huge claw (fig. 134) and other bones in a clay pit within the Wealden Clays. He sensibly notified the Natural History Museum, who excavated all the bones and described a new flesh-eating dinosaur which they christened *Baryonyx* ('heavy clawed') *walkeri*, after the founder. The large claw may have been used for fishing, in the same way that grizzly bears use their claws. This discovery goes to show that amateurs should keep looking and notifying their finds to a local museum. It also adds weight to Darwin's opinion that the missing links are missing because the fossil record is very poor, and as time goes on more and more evidence will be found for new forms of life. We have only one specimen of *Baryonyx* from Britain, but it is impossible to accept that only one ever existed here! If it had not been for the work of John Walker we would

Iguanodon
Height 6 m

Baryonyx
Length 10 m

40 cms

FIG. 134. *Cretaceous dinosaurs,* Iguanodon *and* Baryonyx, *both found in the early Cretaceous rocks of southern England.* Iguanodon *after Black, R.,* Elements of Palaeontology, *Cambridge University Press, 1st edition, 1972, p. 263, fig. 164.*

not even know about the one specimen. Another new dinosaur was discovered in the Wealden Beds of the Isle of Wight in 1998. This is a 4 m long predator, with 10 cm claws, resembling the infamous *Velociraptor* in *Jurassic Park*.

The mass extinction at the end of the Cretaceous

At the end of the Cretaceous 75 per cent of all living species became extinct, but the most famous were the dinosaurs. The ammonites, perhaps the most successful group of marine invertebrates ever, also ceased to exist, as did the giant carnivorous marine reptiles, the

mososaurs, which were up to 15 m in length and weighed up to 40 tonnes. A host of other animals and plants failed to make it into the succeeding Tertiary Period.

The most popular explanation for this mass extinction is the collision with the Earth of a 12 km diameter asteroid or comet. This would have caused huge amounts of dust and debris to pollute the atmosphere and block out sunlight for a considerable period, which would have affected all land animals and plants. Vegetation would have been destroyed, and food chains would have broken down. There would have been nothing for the plant-eating dinosaurs to eat and thus no plant-eating dinosaurs for

Sea-floor spreading in the North Atlantic between Britain and Greenland formed 2500 km of ocean floor during the Tertiary, in two 1250 km wide strips either side of the mid-ocean ridge. This is a spreading rate of 2 cm per year on either side of the ridge giving a total sea-floor spreading rate of 4 cm per year. The rate was similar during the Cretaceous in the Atlantic off Spain. This is a much slower rate than that in the North Pacific, of 3.5 cm (total 7 cm) per year during the Tertiary, and much less than that of the North Pacific during the Cretaceous – 10 cm (total 20 cm) per year.

British Tertiary igneous and volcanic activity

In Britain igneous and volcanic activity lasted a relatively short time from 63 Ma to 52 Ma, and had thus ceased by the early Eocene. In fact most activity had ceased by 57 Ma and the famous basalt lavas of Skye, Mull and Antrim, up to 2 km thick, had nearly all been erupted by 59 Ma. The main phase of eruptions was thus short-lived, covering a time span of some four million years. However in Greenland 4 km of lavas were erupted in only one million years.

The igneous activity was localised at a number of centres (fig. 145) and falls into three main types: basalt lavas, intrusion complexes and dyke swarms.

Great thicknesses of essentially horizontal flows of plateau basalts were erupted from fissure volcanoes in the established Mesozoic basins. Deep-seated plutonic igneous complexes, essentially the roots of old volcanoes now exposed by erosion, formed along fault lines. These include gabbros and granites associated with smaller scale intrusions including dykes, sills, cone sheets and ring dykes, as well as lavas and ashes.

In many places the roots of the old volcanoes and magma chambers have been exposed by erosion and have allowed classic work to be carried out on the details of volcanic and igneous processes. This is particularly true on Skye (fig. 146), Mull, Rhum and Ardnamurchan.

The plateau basalts

On Skye and Mull in particular, as well as the Antrim plateau in Northern Ireland, up to 2000 m of basalt lavas were erupted, compared with 6000 m in Greenland. Lavas vary from 1 m to 50 m in thickness but most are around 10–15 m. On northern Skye they occupy 1000 km^2 and are 600 m thick, and on Mull they cover an area of 700 km^2 and are 1800 m thick. They also occur in western Morvern and at many other of the major centres (fig. 145). On northern and western Skye the skyline is dominated by the stepped or 'trap' landscape created by the lava flows (fig. 147).

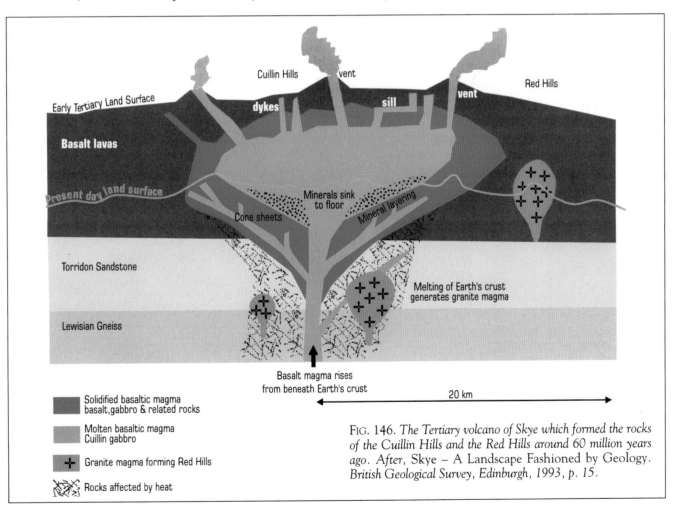

FIG. 146. *The Tertiary volcano of Skye which formed the rocks of the Cuillin Hills and the Red Hills around 60 million years ago. After, Skye – A Landscape Fashioned by Geology. British Geological Survey, Edinburgh, 1993, p. 15.*

The stepped effect is caused by the fact that each successive lava flow has a vesicular top full of gas bubble holes and weathers back more easily. These are horizontal or gently inclined and form distinctive flat-topped hills with stepped sides, as at MacLeod's Tables in Duirinish. The lavas occur on the sea bed across to Northern Ireland, where they form the spectacular Giant's Causeway with its hexagonal basalt columns. These are also found at Fingal's Cave on the Isle of Staffa off the western coast of Mull (fig. 148). The six-sided columns are the result of the perfect cooling of a basalt lava. As it cooled and hardened, cracks or joints formed in a perfect mathematical pattern. Only the central part of the flow shows these columns. The tops of the lavas are usually broken and slaggy with gas bubble holes (vesicles) filled with minerals (amygdales). Many basalt lavas and sills of various ages show these structures but none so perfect as those of Fingal's Cave and the Giant's Causeway.

The lavas were erupted on land from fissures similar to those found in Iceland today. Between episodes of eruptions red laterite (iron-rich) soils developed on top of the flows and these contain plant remains of Tertiary age. Some volcanoes had collapsed calderas in which shallow lakes formed, with tropical plants such as Lotus flowers, indicating a warm, humid climate. Rare pillow lavas indicate eruption into these shallow lakes. The pre-lava landscape was one of hills with river valleys containing coarse sediments. On the Isle of Eigg a thick toffee-like mass of pitchstone lava filled a steep valley and forms the Sgurr of Eigg. Pitchstone is a lava which cooled so quickly that it is almost a glass, similar to obsidian, with little crystal structure. It breaks with a glassy fracture with sharp edges.

Central igneous complexes

These are dominated by large, coarse grained, plutonic masses of gabbro (see below), ultra-basic rocks and granites, surrounded by arcuate suites of smaller intrusions, ring dykes, cone sheets (see below), etc. The ultra-basic rocks, usually peridotites, and the gabbros often show remarkable layered structures. This layering is caused by crystals of different melting points and densities settling out by gravity during crystallisation of the magma. This remarkable feature gives the igneous rock an almost sedimentary appearance (fig. 149). The gabbros and ultra-basic rocks represent magma chambers formed up to 4 km down and since exposed by erosion. They are associated with volcanic vents and collapsed calderas representing surface volcanicity which occurred early in the development of the centre. This early volcanic activity was followed by major intrusions, both shallow (hypabyssal) and deep seated (plutonic). The formation of granites in these centres, as on Skye and Arran, may be due to melting of the basement of Lewisian and Torridonian rocks. The emplacement of the granites sometimes caused disturbance of the

FIG. 147. *Tertiary basalt lavas forming the Storr of Skye. 600 m of lava flows form the well-known stepped landscape of north-east Skye.*